MATHEMATICS EDUCATION IN THE DIGITAL AGE

The wide availability of digital educational resources for mathematics teaching and learning is indisputable, with some notable genres of technologies having evolved, such as graphing calculators, dynamic graphing, dynamic geometry and data visualization tools. But what does this mean for teachers of mathematics, and how do their roles evolve within this digital landscape? This essential book offers an international perspective to help bridge theory and practice, including coverage of networking theories, curriculum design, task implementation, online resources and assessment.

Mathematics Education in the Digital Age details the impacts this digital age has, and will continue to have, on the parallel aspects of learning and teaching mathematics within formal education systems and settings. Written by a group of international authors, the chapters address the following themes:

- Mathematics teacher education and professional development
- Mathematics curriculum development and task design
- Assessment of mathematics
- Theoretical perspectives and methodologies/approaches for researching mathematics education in the digital age

This book highlights not only the complex nature of the field but also the advancements in theoretical and practical knowledge that is enabling the mathematics education community to continue to learn in this increasingly digital age. It is an essential read for all mathematics teacher educators and master teachers.

Alison Clark-Wilson is Principal Research Fellow at UCL Institute of Education, London, United Kingdom. Alison's research spans aspects of designing, implementing and evaluating digital technologies for mathematics education. She leads the European Society for Research in Mathematics Education's working group on "teaching mathematics with technology and other resources".

Ana Donevska-Todorova earned her PhD at the Institute of Mathematics at the Humboldt-University of Berlin and now works in the research field of meaningful design, application and dissemination of digital resources for teaching and learning mathematics. She is fully engaged with the activities of the European Society for Research in Mathematics Education as a member of its International Programme Committee and as a co-leader of the ERME Topic Conferences Mathematics Education in the Digital Age.

Eleonora Faggiano is Assistant Professor in Mathematics Education at the University of Bari Aldo Moro, Italy. Her research interests concern the integration of digital and non-digital resources in teaching practices and mathematics teacher professional development.

Jana Trgalová is Associate Professor at Claude Bernard University, France. She is involved in primary and secondary mathematics teacher education and her research focuses on digital technology and resources for mathematics education, their design, appropriation, use and evaluation.

Hans-Georg Weigand is Professor at the University of Würzburg, Germany. He is interested in the use of digital technologies in mathematics education and teacher education, having authored books on algebra, geometry and computers in mathematics education. He leads the European Society for Research in Mathematics Education's working group on "learning mathematics with technology and other resources".

New Perspectives on Research in Mathematics Education – ERME series

ERME, the European Society for Research in Mathematics Education, is a growing society of about 900 researchers from all over Europe and beyond whose mathematics education research is shaped by a spirit of communication, cooperation, and collaboration.

The contributions in the ERME Series seek to understand and improve learning and teaching of mathematics at schools, colleges and universities, as well as in informal settings (e.g., related to street mathematics or to self-organized networks of teachers). The ERME Series puts an emphasis on reflecting the institutional, societal, and cultural contexts of learners, teachers and researchers and how this context shapes research and adopts a variety of perspectives on its research field.

The volumes are written by and for European researchers, but also by and for researchers from all over the world. An international advisory board guarantees that ERME stays globally connected. A rigorous and constructive review procedure guarantees a high quality of the series.

Volumes of the series:

(5) Research and Development in University Mathematics Education
Overview Produced by the International Network for Research in Didactics of University Mathematics
Edited by Viviane Durand-Guerrier, Reinhard Hochmuth, Elena Nardi and Carl Winsløw

(6) Mathematics Education in the Digital Age
Learning, Practice and Theory
Edited by Alison Clark-Wilson, Ana Donevska-Todorova, Eleonora Faggiano, Jana Trgalova and Hans Georg Weigand

For more information about this series, please visit: https://www.routledge.com/European-Research-in-Mathematics-Education/book-series/ERME

MATHEMATICS EDUCATION IN THE DIGITAL AGE

Learning, Practice and Theory

Edited by Alison Clark-Wilson, Ana Donevska-Todorova, Eleonora Faggiano, Jana Trgalová and Hans-Georg Weigand

Routledge
Taylor & Francis Group
LONDON AND NEW YORK

First published 2021
by Routledge
2 Park Square, Milton Park, Abingdon, Oxon OX14 4RN

and by Routledge
605 Third Avenue, New York, NY 10158

Routledge is an imprint of the Taylor & Francis Group, an informa business

British Library Cataloguing-in-Publication Data
A catalogue record for this book is available from the British Library

Library of Congress Cataloging-in-Publication Data
Title: Mathematics education in the digital age : learning, practice and theory / edited by Alison Clark-Wilson, Ana Donevska-Todorova, Eleonora Faggiano, Jana Trgalova, Hans-Georg Weigand.
Description: Milton Park, Abingdon, Oxon ; New York, NY : Routledge, 2021. | Series: European research in mathematics education | Includes bibliographical references and index. | Identifiers: LCCN 2020057605 | ISBN 9780367684518 (hardback) | ISBN 9780367684525 (paperback) | ISBN 9781003137580 (ebook)
Subjects: LCSH: Mathematics--Study and teaching. | Mathematics--Computer-assisted instruction.
Classification: LCC QA20.C65 M34 2021 | DDC 510.71--dc23
LC record available at https://lccn.loc.gov/2020057605

ISBN: 978-0-367-68451-8 (hbk)
ISBN: 978-0-367-68452-5 (pbk)
ISBN: 978-1-003-13758-0 (ebk)

Typeset in Bembo
by SPi Global, India

CONTENTS

About the Editors ix
Foreword xii
Alison Clark-Wilson

1 Introduction 1
Alison Clark-Wilson, Ana Donevska-Todorova, Eleonora Faggiano, Jana Trgalová and Hans-Georg Weigand

2 Online resources for mathematics teaching and learning at the university level: Three case examples that highlight principles for task design driven by students' and teachers' needs 7
Giovannina Albano, Margo Kondratieva and Agnese Ilaria Telloni

3 Quality of task design in technology-enhanced resources for teaching and learning mathematics 23
Ana Donevska-Todorova, Jana Trgalová, Christof Schreiber and Teresa Rojano

4 Towards pragmatic theories that underpin the design of teacher professional development concerning technology use in school mathematics 42
Eleonora Faggiano, Helena Rocha, Ana Isabel Sacristán and Marisol Santacruz-Rodríguez

5 Technology-rich assessment in mathematics 69
Maria Fahlgren, Mats Brunström, Frederik Dilling, Bjarnheiður Kristinsdóttir, Guido Pinkernell and Hans-Georg Weigand

6 Digital platforms for mathematics teacher curriculum design: Affordances and constraints 84
Ghislaine Gueudet, Birgit Pepin, Scott Courtney, Zeger-Jan Kock, Morten Misfeldt and Andreas Lindenskov Tamborg

7 CAS from an assessment point of view: Challenges and potentials 99
Uffe Thomas Jankvist, Jonas Dreyøe, Eirini Geraniou, Hans-Georg Weigand and Morten Misfeldt

8 Digital maps of the connections in school mathematics: Three projects to enhance teaching and learning 121
Martha Koch, Jere Confrey, Alison Clark-Wilson, Ellen Jameson and Christine Suurtamm

9 The role of technology in the pragmatic-abstract continuum in mathematics curriculum development and task design 138
Allen Yuk Lun Leung and Ana Donevska-Todorova

10 A framework for analyzing students' learning of function at the upper-secondary level: Connected working spaces and abstraction in context 150
Giorgos Psycharis, Georgios-Ignatios Kafetzopoulos and Jean-Baptiste Lagrange

11 Challenges of making sense of tasks and automated feedback in digital mathematics textbooks 168
Sebastian Rezat, Florian Schacht and Uta Häsel-Weide

12 Technology integration for mathematics education in developing countries, with a focus on India and Mexico 185
Ana Isabel Sacristán, Jeenath Rahaman, Suchismita Srinivas and Teresa Rojano

13 Aligning teaching with current experiences of being, becoming and belonging: An identity perspective on the use of digital resources 213
Charlotte Krog Skott, Giorgos Psycharis and Jeppe Skott

Index *228*

ABOUT THE EDITORS

Alison Clark-Wilson, Principal Research Fellow

Alison originally qualified as a chemical engineer before training as a secondary mathematics teacher in the early 1990s. Her interest in digital technologies resulted in her becoming a teacher-researcher in the first UK pilots of mobile technologies and an advanced skills teacher prior to moving to the university sector as a teacher-educator. Subsequently, her research has focused on the design-based research of innovative curriculum resources for secondary mathematics, working in collaboration with industry partners (RM Maths Alive!, Intel Maths Toolkit, Cornerstone Maths, ScratchMaths). She achieved her master's degree in mathematics in 2000 from the University of Chichester and a doctorate in mathematics education in 2010 from the Institute of Education, University of London, now UCL Institute of Education, where she currently works. Her research interests concern broader aspects of scaling and sustaining technology-based innovations in mathematics education and the design and evaluation of associated teacher professional learning initiatives. Alison is a fellow of the Institute of Mathematics and Its Applications (FIMA) and a Chartered Mathematics Teacher (CMathTeach).

Hans-Georg Weigand, Senior Professor

Hans-Georg is a professor for mathematics education at the University of Würzburg (Germany) since 2000. He studied mathematics and physics, and he taught at a German grammar school (Gymnasium). His academic research is in the area of new technologies in mathematics education. He has been at the universities of Eichstätt, Oldenburg and Gießen, and he has been a visiting professor at the University of Illinois (United States). He is involved in pre-service and in-service teacher education. For many years, he has been interested in the use of digital technologies in mathematics education and teacher education. He has written books about algebra, geometry and computers in mathematics education. He was the chair of the technology group of CERME in the last few years, and he was and is the chair of the conference Mathematics Education in the Digital Age.

Ana Donevska-Todorova, PhD, M.Ed.

Ana has qualified as a mathematics teacher for upper secondary school and obtained her master's degree in mathematics education in 2009 at the Ss. Cyril and Methodius University in Skopje. She has received numerous awards for her dedicated involvement in diverse activities related to national and international school mathematics Olympiads. She achieved the title doctor rerum naturalium at the Institute of Mathematics, Humboldt University of Berlin in 2017. For her outstanding design-based research, she obtained a prestigious scholarship, the Research Grant for Doctoral Candidates from the German Academic Exchange Service, and the Caroline von Humboldt Scholarship for excellence. She worked as a research assistant and a research coordinator of mathematics education. Her current research interests relate to mathematics teacher education and teacher development programs about sustainable integration and dissemination of novel digital media and practices. Between 2019-20, her professional occupation was professor in substitution at the Goethe University Frankfurt am Main.

Eleonora Faggiano, PhD, MS, Assistant Professor

Eleonora obtained her master's degree in mathematics in 1999 from the University of Bari and in 2001 qualified as a mathematics and physics secondary teacher from Scuole di Specializzazione per l'Insegnamento Secondario-Puglia. Her interest in the use of digital technologies in mathematics education brought her to achieve the title of doctor of philosophy in computer science in 2004, defending a thesis on mathematics educational technology entitled "A Framework for Supporting Web-Based Cooperative Learning". Her initial works, in the early 2000s, concerned the study of how the use of the emerging computer-supported cooperative learning technologies could foster the students' mathematical learning. Her interest was then attracted by deeper studies concerning dynamic geometry and its implications in the teaching-learning practices. In particular, more recently, she was co-author of papers investigating the synergy between digital technologies and other non-digital resources in promoting students' construction of mathematical meanings and the role of the teacher in the technology-rich classroom. She is currently assistant professor at the University of Bari Aldo Moro where she teaches mathematics and mathematics education.

Jana Trgalová, PhD, Associate Professor

Jana has been an associate professor at Claude Bernard University in Lyon, France, since 2003. She obtained her master's degrees in mathematics and computer science in 1990 from Comenius University, Bratislava, Slovakia, and in mathematics education in 1991 from Joseph Fourier University, Grenoble, France, and her PhD degree in mathematics education in 1995 from the same university. Her research focuses on digital technology and resources for mathematics education, their design, appropriation, use and evaluation. She has participated in several national and international research projects pertaining to digital technologies and resources in mathematics education (e.g., ReMath or MC Squared European projects, respectively, investigating semiotic issues in digital learning environments and focusing on the collaborative design of digital resources fostering mathematical creative thinking). In her current position, she is also involved in primary and secondary mathematics teacher education.

FOREWORD

Alison Clark-Wilson

This edited volume comprises a collection of 12 peer-reviewed chapters that were initiated through collaboration, cooperation and communication during the Fifth Topic Conference of the European Society for Research in Mathematics Education on Mathematics Education in the Digital Age.

FIGURE 1.1 (From left to right) Niels Grønbæk, Alison Clark-Wilson, Eleonora Faggiano, Ana Donevska-Todorova, Hans-Georg Weigand and Jana Trgalová

FIGURE 1.2 Conference participants

This conference took place in Copenhagen, Denmark, from 5 to 7 September 2018, organized by Hans-Georg Weigand (chair of the IPC), Alison Clark-Wilson, Ana Donevska-Todorova, Eleonora Faggiano and Jana Trgalová. Niels Grønbæk was the chair of the LOC.

The conference proceedings of the peer-reviewed research paper and poster presentations are published at https://hal.archives-ouvertes.fr/hal-02500112/.

One of the goals of the conference was to provide substantial time for the nearly 100 participants to not only present and discuss their own work but also to actively seek opportunities to pursue further collaborative research work in relation to the three themes of the original conference, which were as follows:

Theme 1: *Mathematics teacher education and professional development in the digital age*, which focuses on how the digital world has impacted mathematics teacher education (pre-service and in-service); professional development and teachers' professional growth; teachers' professional development practices, collaboration and communities of practice; models and programs of professional development (contents, methods and impacts); and the professional development of teacher educators and academic researchers.

Theme 2: *Mathematics curriculum development and task design in the digital age*, which concerns issues related to digital curriculum materials, resources (including those using digital technology) and e-textbooks, with a focus on their design, appropriation, use and wider dissemination. A sub-theme of *assessment* emerged from the contributions to this theme.

Theme 3: *Theoretical perspectives and methodologies/approaches for researching mathematics education in the digital age*, which addresses how different theories shape research in this field and how they can possibly be combined in a synergic way to address the complexity of teaching and learning processes with digital technology by providing a particular insight into the relationship between uses of technology and the development of students' mathematical knowledge, the role of teachers in a technology-rich environment and teachers' professional development needs, for example, concerning task design.

During the working discussions and built-in 'pause' sessions, the participants were encouraged to examine the particular ways that digital technology is affecting the different knowledge domains within mathematics education research and practice – in particular, how the discerning use of technology requires a deeper understanding of how the mathematics shapes and is shaped by the technology, prompting a rethinking of curriculum hierarchies and closer examination of the relationship between technological and non-technological approaches. In addition, the current (and future) roles of the mathematics teacher in all educational settings require continuous thought as the world grapples with the need for teachers to *both* stay abreast of changes and *simultaneously* lead the way by adapting and developing new pedagogies.

The conference participants were invited to foster new collaborations during the event to enable them to propose jointly written chapters that addressed the new themes that emerged during the conference. The chapter ideas were seeded by the aims to acknowledge the undoubted influences that the digital world is bearing on the teaching and learning of mathematics across all phases of education and to enable researchers from diverse geographical contexts and experiences to work together on theoretical, methodological and practical implications.

In particular, the following themes (and their inter-relationships) emerged at the conference:

- Mathematics teacher education and professional development in the digital age
- Mathematics curriculum development and task design in the digital age and more specifically the design principles underpinning digital curricula, mathematical tasks in technology-enhanced environments and e-books for mathematics
- Particular design and implementation of assessments/examination, to include e-assessment
- Theoretical perspectives and methodologies/approaches for researching mathematics education in the digital age

In the 18 months following the conference, 43 authors cooperated, collaborated and communicated with each other to produce the 12 edited chapters that are presented here.

Giovannina Albano, University of Salerno, Italy
Mats Brunström, Karlstad University, Sweden

Alison Clark-Wilson, University College London, UK
Jere Confrey, North Carolina State University, United States
Scott Courtney, Kent State University, United States
Frederik Dilling, Universität Siegen, Sweden
Ana Donevska-Todorova, Goethe University Frankfurt, Germany
Jonas Dreyøe, Aalborg University, Denmark
Eleonora Faggiano, University of Bari Aldo Moro, Italy
Maria Fahlgren, Karlstad University, Sweden
Eirini Geraniou, University College London, UK
Ghislaine Gueudet, University of Brest, France
Uta Häsel-Weide, Paderborn University, Germany
Ellen Jameson, University of Cambridge, UK
Uffe Thomas Jankvist, Aarhus University, Denmark
Georgios-Ignatios Kafetzopoulos, National and Kapodistrian University of Athens, Greece
Durgaprasad Karnam, Homi Bhabha Centre for Science Education, TIFR-Mumbai, India
Martha Koch, University of Manitoba, Canada
Zeger-Jan Kock, Eindhoven University of Technology, Netherlands
Margo Kondratieva, Memorial University, Canada
Bjarnheiður Kristinsdóttir, University of Iceland, Iceland
Charlotte Krog Skott, University College Copenhagen, Denmark
Jean-Baptiste Lagrange, Université Paris Diderot, France
Allen Yuk Lun Leung, Hong Kong Baptist University, Hong Kong
Morten Misfeldt, Aalborg University, Denmark
Birgit Pepin, Eindhoven University of Technology, Netherlands
Guido Pinkernell, Pädagogische Hochschule Heidelberg, Germany
Giorgos Psycharis, National and Kapodistrian University of Athens, Greece
Jeenath Rahaman, Tata Institute of Social Science, India
Sebastian Rezat, Paderborn University, Germany
Helena Rocha, Universidade Nova de Lisboa, Portugal
Teresa Rojano, Centre for Research and Advanced Studies, Mexico
Ana Isabel Sacristán, Center for Research and Advanced Studies, Mexico
Marisol Santacruz-Rodríguez, Universidad del Valle, Colombia
Florian Schacht, University of Duisburg-Essen, Germany
Christof Schreiber, Justus-Liebig-Universität Giessen, Germany
Jeppe Skott, Linneus University, Sweden
Suchismita Srinivas, Tata Institute of Social Science, India
Christine Suurtamm, University of Ottawa, Canada
Andreas Lindenskov Tamborg, Aalborg University, Denmark
Agnese Ilaria Telloni, University of Ancona, Italy
Jana Trgalová, Claude Bernard University - Lyon 1, France
Hans-Georg Weigand, University of Würzburg, Germany

1

INTRODUCTION

Alison Clark-Wilson,

UNIVERSITY COLLEGE LONDON, UK

Ana Donevska-Todorova,

GOETHE UNIVERSITY FRANKFURT, GERMANY

Eleonora Faggiano,

UNIVERSITY OF BARI ALDO MORO, ITALY

Jana Trgalová and

CLAUDE BERNARD UNIVERSITY, FRANCE

Hans-Georg Weigand

UNIVERSITY OF WÜRZBURG, GERMANY

First and foremost, the topic *Mathematics Education in the Digital Age* encompasses the whole field and any single edited volume on the subject cannot possibly do justice to the cumulative knowledge of over 50 years of mathematics education research. Consequently, we aim to privilege the context of the *digital age* by offering a comprehensive text that highlights the impacts this has, and will continue to have, on the aspects of learning and teaching mathematics in formal education systems and settings. In this editorial, we focus one by one on the four main themes addressed within the book, which are as follows:

- Mathematics teacher education and professional development in the digital age
- Mathematics curriculum development and task design in the digital age
- Assessment of mathematics in the digital age
- Theoretical perspectives and methodologies/approaches for researching mathematics education in the digital age

We highlight how each of these themes is addressed by the chapter authors, either explicitly or more implicitly, to draw out the important interconnections within and between the respective texts. By doing so, we aim to highlight not only the complex nature of the field but also the advancements in theoretical and practical knowledge that is enabling the mathematics education community to continue to learn in this increasingly digital age.

Mathematics teacher education and professional development in the digital age

Since the 1980s and 1990s and the introduction of the early mathematical computer applications to mathematics classrooms (i.e. LOGO, spreadsheets, dynamic graphing and dynamic geometry software), teachers of mathematics have been challenged to explore how these new tools impact on the curriculum and its pedagogy. In the intervening years, the mathematics education research field has expanded to consider the teacher in all of his/her roles within the context of both pre-service and in-service teacher education, with increasing emphasis on how to prepare and support teachers to conduct their professional lives in this fast-moving digital age.

The chapter by *Faggiano, Rocha, Sacristán and Santacruz-Rodríguez* adopts a historical approach to the analysis of major national or regional professional development initiatives in Colombia, Italy, Mexico and Portugal to aim at making explicit the "pragmatic theories" that have underpinned their design, implementation and (in some cases) evaluation. Their resulting methodological frame develops and expands five interconnected themes: the vision and goals of the initiative, its focus, the strategies and methods employed, constraints and incentives and the approaches to evaluation. This frame offers a useful analysis tool to inform the design of teacher professional development interventions that address digital technology use in school mathematics education – and take account of important international differences.

Sacristán, *Rahaman*, *Srinivas* and *Rojano* provide the important perspectives from developing countries in their chapter, which offers insights from Mexico and India on technology integration in mathematics education. Within this, they consider the contextual factors that influence teachers' attitudes to and uptake of technology, alongside the nature of associated professional development design.

Gueudet, *Pepin*, *Courtney*, *Kock*, *Misfeldt* and *Lindenskov Tamborg* also adopt an international perspective as they compare the affordances and constraints of digital education platforms offered as resource banks for teachers of mathematics to support them to design lessons. Drawing on the documentational approach, the authors conclude a number of factors that differentiate such platforms, which link to national education policies and perspectives on the nature of teachers' work, in particular the important aspect of the role of the teacher as a designer of lessons. Their chapter highlights the developing complexity of this role in the digital age.

Aspects of the role of the teacher are addressed less explicitly in the chapter by *Jankvist*, *Dreyøe*, *Geraniou*, *Weigand* and *Misfeldt* (concerning teachers' roles as formative assessors of students' learning), that by *Albano*, *Kondratieva* and *Telloni* (reflecting the multiple roles of undergraduate mathematics' lecturers as designers, teachers and researchers of students' digital experiences) and the chapter by *Leung* and *Donevska-Todorova* (offering pedagogical insights into how the pragmatic to abstract continuum is addressed within digital task design).

Finally, three chapters produce tools and frameworks that may prove useful for teachers in aspects of their roles as curriculum and lesson designers.

Donevska-Todorova, Trgalova, Schreiber and *Rojano* compare the analysis of selected digital mathematics tasks to develop a set of criteria that might indicate a sense of their quality.

Koch, Confrey, Clark-Wilson, Jameson and *Suurtamm* describe and analyze three digital maps of the mathematics curriculum, which offer different ways to conceive its design and construction. The resulting maps offer opportunities to reflect on the design and enactment of the mathematics curriculum in dynamic ways.

Skott, C., Psycharis and Skott, J. adopt a social perspective to help understand how the use of digital resources affects one teacher's professional by using the *Patterns of Participation* (PoP) framework.

Research on aspects of initial teacher education and teachers' ongoing professional learning is evolving rapidly to incorporate aspects of digital technological tools beyond only training activities with particular digital tools. However, the geographical and cultural diversity of technology integration within mathematics education, as brought to the fore by several chapters in this book, indicates that the dissemination of existing research, with further validation of the theories and methods adopted, may support more research-informed approaches. However, as new mathematical technological tools emerge, alongside the increased interoperability of the existing tools, there will always be the continued need for related research. There is still much to be understood about the components of more effective teaching practices involving technological tools and how these relate to the evolution of teachers' knowledge.

Mathematics curriculum development and task design in the digital age

This theme concerns issues related to digital curriculum materials, resources (including those using digital technology) and e-textbooks with a focus on their design, appropriation, use and wider dissemination.

Task design is explicitly addressed in chapters by *Albano, Kondratieva* and *Telloni*; *Leung* and *Donevska-Todorova*; and that by *Donevska-Todorova, Trgalova, Schreiber* and *Rojano*.

The chapter by *Leung* and *Donevska-Todorova* explores the potential of digital technology to allow students to experience mathematical ideas and concepts in learning situations where discursive mathematical practices supported by digital artifacts gradually evolve into a formal mathematical knowledge pragmatic-abstract continuum. They call for designing technology- based tasks addressing the *pragmatic-abstract continuum*, which they illustrate in examples from primary to university levels.

Albano et al. suggest using a tetrahedron model, considering the didactic system in the e-learning context composed of a mathematical knowledge to be taught, a learner, a tutor and an author to design tasks for blended learning. The authors present and discuss examples of online resources to be used in face-to-face learning at the university.

Donevska-Todorova et al. question the quality of digital resources (e.g., apps or audio podcasts) and technology-based mathematical tasks (e.g., including dynamic geometry or spreadsheet). The authors suggest a set of criteria that might help teachers select the most adequate ones or serve as guidelines in task design.

The use of existing digital resources and their impact on students' learning is dealt with in the chapter by *Jankvist, Dreyøe, Geraniou, Weigand* and that by *Misfeldt* and *Rezat, Schacht* and *Häsel-Weide.*

Rezat et al. focus on students' interactions with digital textbooks and highlight challenges students face with two characteristic features of these digital resources – namely, interactive answering formats and automated feedback. Implications the authors draw from their study point to the need of a careful use of such a resource in the classroom, which requires teacher's awareness of the goals of the tasks and of possible effects of automated feedback on students' mathematical activity.

Jankvist et al. analyze Computer Algebra System (CAS)-supported assessment tasks to highlight benefits from using technology that allows for formative assessment or self-assessment and that in some cases shifts the focus to mathematical concepts rather than to complex calculations. On the other hand, the authors show examples in which technology cuts away the relevance of the mathematical task.

The chapter by *Gueudet, Pepin, Courtney, Kock, Misfeldt* and *Lindenskov Tamborg* focuses on digital educational platforms available for mathematics teachers. The authors analyze affordances and constraints of the platforms as tools supporting teacher curriculum design.

In their chapter, *Koch, Confrey, Clark-Wilson, Jameson* and *Suurtamm* present three initiatives of mathematics curriculum digital mapping. The three different approaches share a common goal to enhance mathematics teaching and learning by offering another view of the curriculum conveying connections between mathematical concepts and providing access to related instructional resources or assessment tasks.

These contributions open new avenues for further research. Interactivity appears as the foremost affordance of technology-supported tasks, and feedback is, therefore, a critical issue in task design. It requires a deep didactic and cognitive analysis. Personalized and adaptive feedback, feedback helping students become autonomous learners or scaffolding that gradually fades away are some of the task design features calling for future developments. Further research is also needed to explore teachers' use of new resources, such as educational platforms, e-textbooks, digital curricular maps or resource quality criteria, and their effects of mathematical teaching and learning.

Assessment of mathematics in the digital age

Evolution in curriculum and task design influenced by the use of digital technologies has to mirror in evaluation and assessment in mathematics education. The quantity and diversity of approaches and tools for this aim evolve together with concerns about access, individualization and securing efficacy in teaching and learning.

In this sense, *Jankvist, Dreyøe, Geraniou, Weigand* and *Misfeldt* discuss the use of CAS for the purpose of summative and formative assessment and the ways in which CAS alters or augments these. They also offer examples of examination tasks from secondary school, upper secondary school and university level in Denmark, Germany and the United Kingdom, which have been applied in different contexts. As a result, the authors conclude that CAS is a challenge to the traditional assessment scheme and allows for new types of assessment situations to be implemented because of the increased computational power that CAS provides. However, there is on the one hand the danger that these new types are too complex and too difficult. On the other hand, these new types could end up promoting pragmatic strategies for solving standard tasks with CAS due to such training being aligned with the assessment. There seems to be no general solution for summative assessment. Maybe mixed examinations, one part of the examinations with technology, the other part without, might be a good compromise. Another possibility is emphasizing formative assessment. This type of assessment with digital technologies can support both the students and the teacher. Teachers will get summative results on students' performance, and this information enables teachers to make informed decisions about next steps for their students.

Fahlgren, *Brunström*, *Dilling*, *Kristinsdóttir*, *Pinkernell* and *Weigand* set out a framework for analyzing technology-rich assessment systems, especially for formative assessment. This framework is about assessment through and with technology, the change of the content and the functionality and aspects of feedback. It is used to analyze three distinctive and innovative formative assessment systems: STEP (Seeing the Entire Picture) from Israel, SMART (Specific Mathematics Assessments That Reveal Thinking) from Australia and STACK (System for Teaching and Assessment Using a Computer Algebra Kernel) from the United Kingdom. As a result, the framework allows a rough comparison of the systems; however, a more fine-graded framework is necessary to pinpoint some characteristics and differences between technology-rich assessment systems. One important conclusion of this analysis is the guideline that formative assessment has to be integrated into a learning process, and the use of digital assessment systems might be a part of this process.

Interesting questions about automated feedback for self-assessment can be brought into connection with digital textbooks. For example, *Rezat*, *Schacht* and *Häsel-Weide* discuss feedback in relation to student's activation when engaging with digital textbooks.

Stepping outside of a comfort zone of already known practices and taking responsibility for changes when applying digital media in mathematics education is challenging, in particular in the period of a pandemic crisis. Such changes are required not only in the teaching methods, techniques and tools but also in assessment. This is an intensively growing issue, which is to be adequately addressed during the follow-up European Research in Mathematics Education Group's Topic Conference on Mathematics Education in the Digital Age (MEDA 2), opening many new questions for further research.

Theoretical perspectives and methodologies/approaches for researching mathematics education in the digital age

Research in mathematics teaching and learning, especially in the digital age, deals with complex phenomena involving multiple theoretical perspectives and different kinds of methodological approaches. The chapters in this book refer to many different theories – e.g., design theories, prescriptive theories, theories linking research and practice, theories addressing the transfer of learning arrangements to other learning conditions – that eventually, combined/networked in a synergic way, shape research in mathematics education in the digital age.

For example, the chapter by *Gueudet*, *Pepin*, *Courtney*, *Kock*, *Misfeldt* and *Tamborg* concerning the impact of digital platforms in terms of teachers' curriculum design as aspect/s of teachers' documentation work, combines specific theoretical tools developed within the Documentational Approach to Didactics (the instrumentation and instrumentalization processes and the concept of connectivity) to investigate the affordances and constraints of digital platforms in three different national contexts.

Aspects of theoretical perspectives and methodological approaches are explicitly addressed in the chapters by *Faggiano et al.*, *Psycharis et al.* and *Skott et al.*

Faggiano, *Rocha*, *Sacristán* and *Santacruz-Rodríguez* offer an empirically informed and research-based approach to support the design of mathematics teachers' professional development to exploit technological affordances and prepare mathematics teachers for present and future digitally enhanced contexts. By exploring the evolution of and implications from mainly large-scale national professional development initiatives, they aim to arrive at a more theoretical grounding for the topic, supported by case examples that exemplify the most pertinent aspects. To do this, they have identified a set of aspects (and sub-aspects) that emerged in relation to five major themes and reveal a set of 'pragmatic theories', alongside a consideration of the interconnections between these aspects.

Psycharis, *Kafetzopoulos* and *Lagrange* present a framework for analyzing classroom situations by networking Connected Working Spaces (CWS) and Abstraction in Context (AiC). Their perspective allows a deeper look at students' cognitive evolution as they experience functions in a plurality of digital and non-digital settings (physical context, geometry, measures, algebra). Furthermore, they give insight into the relationship between the use of technology and the development of students' mathematical knowledge. With the attempt to focus on how the combination of AiC and CWS can help to make sense of students' conceptualization of function as covariation, they also underline that the networked analysis may sensitize task designers, teachers and researchers to the complexity of students' conceptualizing processes.

Skott, C., *Psycharis* and *Skott, J.* bring new perspectives and understandings of teacher's development in the use of digital resources. They use a social practice perspective (PoP) to investigate teachers' roles in technology-based learning environments and the influence of culture and society on the way digital resources are used in the classroom. Comparing the findings with a previous study in which a different theoretical approach was used, the authors claim that different frameworks and theoretical approaches – including networking and double analyses – are needed to develop balanced understandings of the integration of digital resources in mathematics classrooms that acknowledge the significance of teachers' professional identities.

2

ONLINE RESOURCES FOR MATHEMATICS TEACHING AND LEARNING AT THE UNIVERSITY LEVEL

Three case examples that highlight principles for task design driven by students' and teachers' needs

Giovannina Albano,

UNIVERSIY OF SALERNO, ITALY

Margo Kondratieva and

MEMORIAL UNIVERSITY, CANADA

Agnese Ilaria Telloni

UNIVERSITY OF ANCONA, ITALY

Introduction

The teaching of mathematics at the university level is affected by several critical issues that cause many students to fail in their study of the subject (see, e.g., De Guzman, Hodgson, Robert, & Villani, 1998). It is evident that merely "exposing students to the course content has often not been enough for them to achieve academic success in mathematics" (Chekour, 2018, p. 50), suggesting that innovations in teaching and learning approaches are required to maintain high-quality university mathematics education. In recent years, research highlights the effectiveness of using e-learning environments to support the teaching and learning processes, mainly at the tertiary level of instruction (Albano & Ferrari, 2008; Albano, 2011; Bardelle & Di Martino, 2012; Kondratieva, 2013). Nardi (2017) notes that there is an "aspiration (institutional but not only) for a more approachable, more inclusive and more engaging learning experience in university mathematics that is tailored to individual student needs" (p. 10) and suggests that e-learning could provide a solution to this aspiration. In particular, blended learning that combines face-to-face (f2f) and online instruction seems most promising (Mundt & Hartmann, 2018). However, current research on how to leverage the potential of blended learning is limited, especially at the university level.

Our research question is, Which elements should be taken into account in designing online tasks that aim to both enhance mathematical learning and scaffold university transitions? This question guides our investigation of the theoretical issues concerning task design in a blended environment at the university level, assuming a "co-disciplinary perspective" (mathematics education, general didactics, technology; Albano, Faggiano, & Rossi, 2018).

Initiated by the analysis of case examples of the design of online resources, the chapter will provide guidelines for effective task design by considering the implications on both students' learning experiences and the nature of the teacher's work.

Education first: The need for a preliminary analysis

The large amount of advanced material to be studied in a relatively short time frame renders it unlikely that students at the university level can construct complete mathematical knowledge without explicit and specifically planned guidance on how to approach the learning of the mathematical content and the use of technology for that purpose.

High-quality task design within the digital environment is required, which takes account of fundamental top-level goals both at the cognitive level (development of formal and relational thinking through argumentation and towards proof) and at the metacognitive and affective levels (engagement, awareness, development of autonomy). In this respect, both the individualization of teaching/learning and cooperation among peers are important, which at the university level are only fully achievable through careful design of interactive environments and tasks. Here we refer to the notion of *individualization* defined by Baldacci (2006) as the differentiation of the didactical paths in order to enable all the students to reach common objectives; this notion is different from *personalization*, which is the differentiation of the formative goals and objectives to promote individual potential.

The tetrahedron model for didactic systems in e-learning (Albano, 2017; Albano, Faggiano, & Rossi, 2018), shown in Figure 2.1, is used to discuss the particular stages of designing for learning. According to this model, the didactic system in e-learning environments is composed of four characteristics: knowledge to be taught/learnt, that is Mathematics (M); someone who is expected to learn M, that is the Student (S); someone who helps S to learn M, that is the Tutor (T); someone who is in charge of the complex design and implementation activity, the Author or Designer (A). The complexity of the technological environment and the need for clear educational purposes suggests that the Author's role is best played by a collective that may include mathematics teachers, experts in mathematics education and in the specific mathematical subject at stake and experts in the use of technology. The "collective" evokes a co-disciplinary perspective for the design: it allows the co-thinking, the integration and the negotiation of the right balance between the various viewpoints (general education, mathematics education, technology) and expertise in the collective focused on the same object of study (Albano, Faggiano, & Rossi, 2018).

The Tutor's role is played by two different actors: (1) the instructor, who meets students during the f2f lessons, and (2) the technology, which can act as a non-human

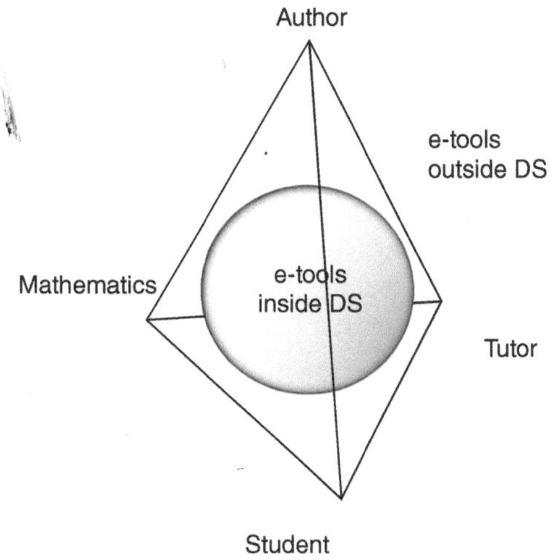

FIGURE 2.1 The tetrahedron model for didactic systems in e-learning (Albano, 2017; Albano, Faggiano, & Rossi, 2018)

tutor (Geiger, 2006) by scaffolding students' interactions with tasks and providing response-specific feedback (Shute, 2008).

Within the tetrahedron model, technology is represented as an inscribed sphere, highlighting that the added-value does not reside in the e-tools themselves but in their exploitation with respect to some specific educational aims (Arzarello, Drijvers & Thomas, 2012).

Putting in background technology, task design in e-environments shows the need for a preliminary analysis along three dimensions: cognitive, epistemological and didactic. It should be followed by an a priori analysis focused on the choices done in designing specific tasks (Artigue & Perrin-Glorian, 1991).

The **cognitive** dimension, in particular the identification of learning difficulties that students encounter, requires the analysis of common misconceptions about the mathematical topic at stake and a consideration of external factors. For instance, many first-year undergraduates experience issues during the transition from high school to university; other students face the transition from the bachelor's degree to the master's degree. The former transition includes noticeable increases in class size and the pace of teaching. This often implies greater diversity among students' level of preparedness for the courses. Also, many students are taught in high school to reproduce given procedural methods rather than to think critically and flexibly and thus any novelty or deviation from standard tasks creates for them impenetrable challenges. This may lead to a weaker understanding of mathematical statements or to assign them imprecise or unintended meanings. Similar phenomena appear during students' transitions to the graduate level. Dealing with both transitions requires students to have higher levels of responsibility and self-efficacy. Often, students' cognitive challenges increase due to a lack of immediate and individualized response to their (mistaken) performance and questions, which prevents the students from proceeding with their studies in a meaningful way.

Students' cognitive difficulties can be observed from the following perspectives of the didactic tetrahedron:

1) In the S-A-T face, S experiences new constraints set by A and T.
2) In the S-M-T face, on the one hand (S-M), students' previous experiences with M greatly affect their current study of M, and on another hand (S-T), students may experience a new form of relationship with their instructor. Here technology can act as a mediator.

Thus, the inscribed sphere comes into play when the Author chooses appropriate e-tools and carefully designs their use to enable the cognitive individualization (S-M) on the one hand and to recover a non-anonymous student-teacher relationship on the other hand (S-T).

The **epistemological** dimension of the preliminary analysis concerns the nature of mathematical knowledge to be taught/learnt. In particular, we analyze (a) the epistemological challenges offered by the mathematical topic and (b) the opportunities provided by a digital environment related to these difficulties. From the A-M-S perspective: A identifies in M worthwhile aspects for S, considers the difficulties and the challenges for S in M and uses technology with the explicit didactic purpose of supporting S in overcoming his/her difficulties and in grasping the richest ideas in M.

The **didactic** dimension of the preliminary analysis is the portrayal of how the teaching of a course is organized: the duration and time schedule, teaching format, assessment, choice of examples and resources and so on. How to use online tasks in a blended setting is an issue concerning the didactic dimension. The following processes can be viewed from the corresponding faces of the didactic tetrahedron: (1) A-T-S: A defines time constraints for a course instructor T and S; (2) A-M-T: A defines M-content constraints for T; (3) T-M-A: T chooses examples, resources and tasks in M under constraints set by A, and T may "modify" M chosen by A; (4) S-M-T: students work on suitable resources in M under supervision of T. We identify the problem occurring within a solely f2f didactical organization: for many students, there is insufficient time for the meaningful processing of given information during the lecture, combined with the minimal internal motivation to do it outside of class. This negatively affects the quality of such students' learning and can often result in the failure of the course. Technology can also play a role at the non-cognitive level by engaging and motivating students to work independently online (Albano & Ferrari, 2008).

Three case examples of the design of e-resources

In this section, we describe three particular online resources designed, implemented and experimented in three projects to support f2f courses in university mathematics departments. The first, *Interactive Texts*, at the Department of Mathematics and

Statistics at Memorial University (Canada). The remaining two *Route of Activities* and *Tree of Online Tasks* at the University of Ancona (Italy). The resources were designed within a formative assessment frame (Black & Wiliam, 2009; Hattie & Timperley, 2007; Cusi, Morselli, & Sabena, 2017) to be useful for both students and teachers since the former could be helped to attain meaningful and conscious learning, whilst the latter could adjust their strategies of teaching in response to the students' reactions. In all cases, the instructors envisaged the possibility of sharing digital resources, but there were no mandatory requirements for students to participate in online activities during the three projects.

The activities described in the following sections assume the construct of hypothetical learning trajectories – "paths by which learning might proceed" (Simon, 1995, p. 135) – devised by the teacher to be performed by students in reaching some learning goals. There are three aspects in a learning trajectory: the learning goal, the advance in thinking and learning which develops progressively and the instructional sequence (learning activity), which is composed of suitable key instructional tasks designed to promote the desired advancement (Clements & Sarama, 2004).

Interactive Texts

Interactive Texts (IT) was devised for use in a precalculus course at Memorial University. The course comprised 48 lecture hours delivered in a 12-week term. The task was inspired by the following problem: *given two curves, and knowing that one of the curves can be transformed into another by an elementary transformation such as translation, reflection or stretch, or a combination of them, identify these transformations*. The preliminary analysis identified the following possible broad issues:

a) the diversity in students' levels of preparedness;
b) somewhat counterintuitive mathematical content (e.g., the graph of the function $y = F(x + k)$, $k > 0$ moves to the left from the graph $y = F(x)$ by k units despite the plus sign in the argument $x + k$ that may be associated with motion to the right) and students' difficulties with the coordination of mathematical terminology, formulas and geometric shapes (Kondratieva & Radu, 2009); and
c) the deficit of lecture time.

IT is a set of written mathematical instructions with interactive diagrams (created in the dynamic geometry software GeoGebra) to enable students to interact with formulas and diagrams. The goal of introducing IT is to enable students to study the topic at their own pace by reading the instructions and solving problems whilst making sense of the emerging constructions by interacting with various interactive diagrams. Initially, no help from the instructor was foreseen: the IT was designed by the author with the aim that students would study the material autonomously or possibly in groups. The text consisted of ten short sections. Most sections were followed by online exercises so that students could assess their own work and obtain immediate feedback not only from experimenting with applets but also from answering embedded multiple-choice questions. After mastering individual transformations, they proceeded to a section titled "Putting It All Together", where they could experiment with given functions and in the later sections introduce their own functions (for more details on the IT resource, see Kondratieva, 2018).

The teaching platform was equipped with e-tools that supported various interactive diagrams (ID) that fit with the following constructs: *illustrating*, *guiding* and *elaborating* (Naftaliev & Yerushalmy, 2017).

> Illustrating IDs demonstrate the objective of the activity to the reader, usually by offering a single representation and relatively simple actions,…an illustrating ID might allow learners to manipulate rather than read a definition. (ibid, p.156)

For instance, in order to establish the connections between the terminology and the visual representation of the *reflection*, *shift* and *stretch* in either vertical or horizontal directions (without references to algebraic notations), students are asked to move the hands of Mark the Mathematician by using the control panel related to functions' transformations (Figure 2.2, left) to achieve one of the positions shown in Figure 2.2.

Next, the focus is on the coordinates of particular individual points. Figure 2.3 (left) shows a point reflected with respect to an axis of symmetry. The learner can

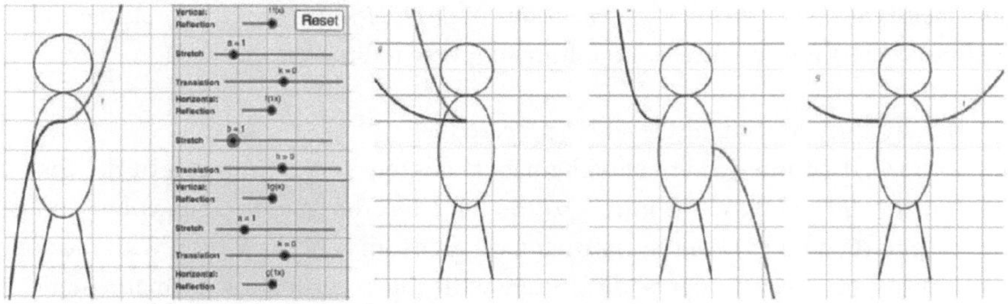

FIGURE 2.2 An entertaining exercise on functions' transformations

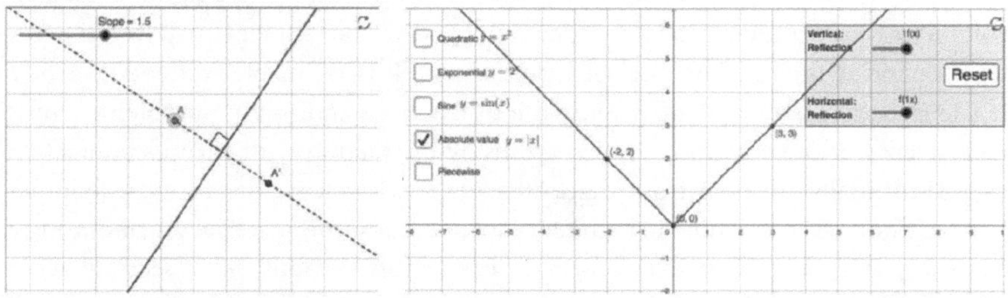

FIGURE 2.3 ID: The axial reflection of a point (left) and of a curve (right)

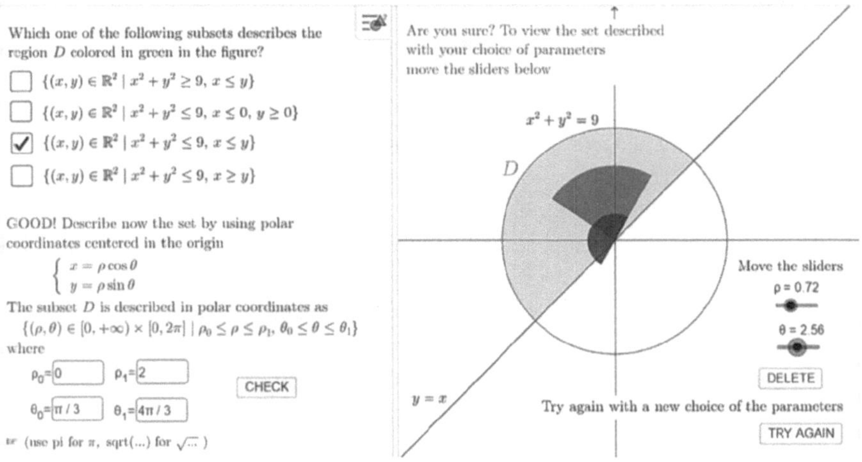

FIGURE 2.4 A task focused on multiple representations of a subset in the plane

For the second part: if the answer is correct, the program signals it, and the student is addressed to the subsequent task; if the answer is wrong, two sliders for ρ and θ appear on the graph and, as the student moves the sliders between the minimum and the maximum values previously inserted, the program draws the corresponding set in the Euclidean plane, thus showing the mistake. These sliders are engineered to allow a direct and dynamic comparison between the representation in polar coordinates and the graph of the region.

In order to promote the learning from a cognitive and a metacognitive point of view, and to avoid a trial-and-error approach, the RA is organized in a methodological-educational cycle structured in the following different phases, which each student faces individually: (P_1) the student receives a question about the recognition of a subset of the Euclidean plane; (P_2) the RA is handed out: the student interacts with the digital environment; (P_3) the student reports his/her reflections on the activities (troubles, increasing difficulties or not); (P_4) the question given in P_1 is submitted again to the student, with the possibility of changing his/her previous answer; (P_5) the student provides his/her thoughts and opinions about the usefulness of the activities to answer correctly the question received in P_1.

Tree of Online Tasks

The Tree of Online Tasks (TT) was designed and implemented for master's degree engineering students attending the course Probability and Mathematical Statistics at the University of Ancona and is focused on elementary and conditional probability (Cusi & Telloni, 2019). The course is taught in the first semester and takes 48 hours of lectures. The preliminary analysis highlights typical students' difficulties in understanding the meaning of conditional probability and the concepts of incompatibility and independence of events. Often, a major source for errors is the coordination between the formal notation and the verbal language for expressing probabilities (Díaz & de la Fuente, 2006). This situation generates "the need for further research on appropriate learning trajectories for students in concepts as conditional probability, independent events, compound probability, and mutually exclusive events" (Cai, 2017, p. 516). The analysis led us to create individualized resources to support both the sequence that each student would take through the tasks and the specific feedback and hints provided for each task.

The tasks within the TT are dynamically connected. This is in tune with the idea that the task sequence of a learning trajectory is not "the only, or the best, path for learning and teaching, only that it is hypothesized to be one fecund route" (Clements & Sarama, 2004, p. 84). Indeed, the various digital tasks of a TT are organized as an intertwined collection of various instructional sequences which constitute individualized learning paths, depending on the students' answers and difficulties during the interaction. The tree includes five tasks implemented in GeoGebra. For each task, the student is required to give open or multiple-choice, numerical or symbolic answers, and he/she can choose to ask for specific hints.

The diagram in Figure 2.5 summarizes the structure of the TT. All learning paths start with the same task E1, in which a text in verbal language is provided and the probabilities of six events are required. Three of them are given in the text, and the other three are obtainable by using the definition of conditional probability and the rules for the probabilities of the complementary event and the intersection event.

When facing task E1, the student can ask for five different hints (see Figure 2.6): a summary of the data given within the text, represented through symbolic expressions (*d*); the Euler-Venn diagrams of the events (*EV*); a calculating machine (*c*); a sheet and a pen (*sp*); and a list of useful formulas (*f*). If a student makes more than

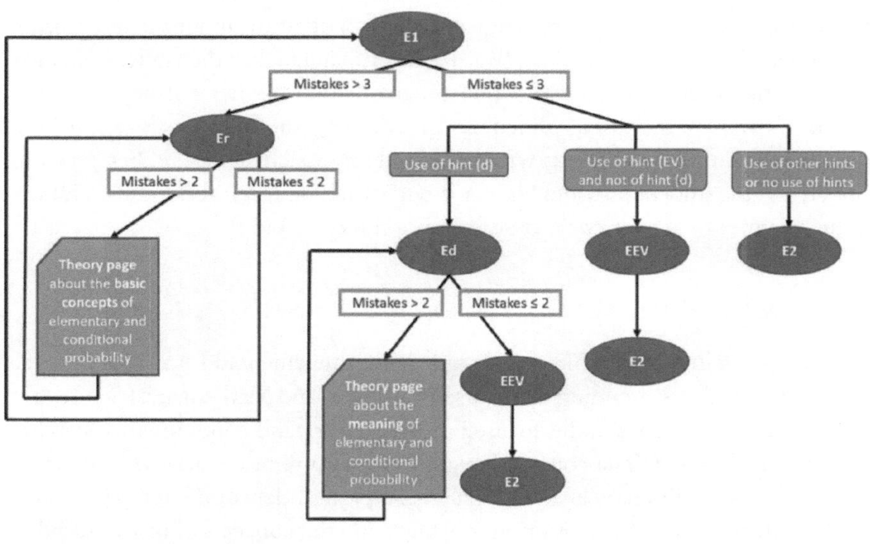

FIGURE 2.5 The tree of tasks

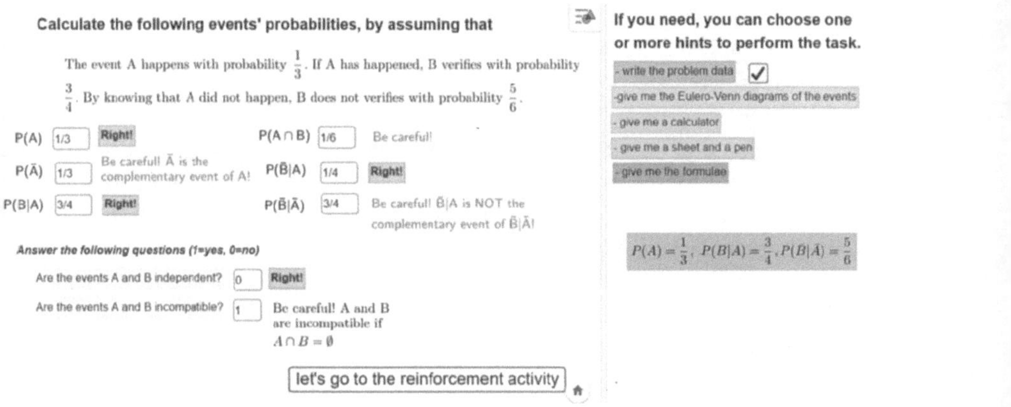

FIGURE 2.6 A screenshot from task E1

three mistakes, he/she is directed to a reinforcement task (*Er*), which is focused on the probability of the complementary of an event and on the definition of conditional probability. In the contrary case, the next task depends on the hints that the student requires: if he/she asks for the hint (*d*), he/she is addressed to a specific task to assess understanding of the given elements in a text expressed in verbal language (task Ed); if he/she does not ask for the hint (*d*), but he/she asks for the hint (*EV*), he/she is addressed to a specific activity about the meaning of events in terms of set operations (task EEV); if the student asks neither for hint (*d*) nor for hint (*EV*), he/she is addressed to task E2, which involves all the knowledge needed for E1, but it is expected to be more difficult since it also requires a challenging part of modelling.

All tasks in the TT provide response-specific feedback designed by taking into account misconceptions and typical mistakes that affect the topic at stake (Díaz & de la Fuente, 2006). This aspect is displayed in Figure 2.6, where examples of response-specific feedback are shown in the boxes on the left side of the screen.

Discussion and guidelines for task design

Our experiences conclude that the design and implementation of online tasks need to go through several trials. This process requires a collaboration of the "collective" Author and Tutor. It relies on the preliminary analyses (A-T-M perspective), alongside students' feedback collected by the Tutor, which involves the interaction of S with the resources designed by the Author (A-T-S perspective). As a result, adjustments are made. Next, we present emerging desired characteristics of tasks suitable for blended learning at the university level.

Individualization. One of the main promising features of online resources is the individualization of teaching and learning. As in the case of the TT, students, after receiving response-specific and formative feedback, are automatically guided to a didactical path depending on their own learning needs. Moreover, in all three cases, students are stimulated through various learning channels (graphical, symbolic, verbal, etc.) and in TT, they can choose hints with different features according to their own preferences.

Effective feedback. It seems that a crucial choice in designing the RA and its implementation was the alternation of immediate and delayed feedback. Within the RA, the software provides immediate feedback when the students submit their responses during the interaction with the tasks in order to scaffold their work and facilitate their decision in the digital environment. Whilst no immediate feedback is provided when the students give their response to the initial question (phase P_1), the same question, along with the answer they gave, is displayed to the students after their interaction with the online tasks (phase P_4). Similarly, in IT, students may return to the initial question that involves a combination of transformations after they become confident regarding each individual transformation. In this way, the students reconsider their first answer with more awareness, possibly changing it or its justification. Hence, they are engaged in cognitive and metacognitive processes and have the possibility to realize the usefulness of the interaction in the digital environment in order to correctly perform the initially given task and gain a sense of self-efficacy since the activity has been done without the presence of a tutor.

Possibility of self-assessment. Students may also use each of the three activities as a self-assessment tool for the following reasons: the activity can be used without the tutor's presence and hence without the fear of being judged, the automatic feedback suggests which theoretical ideas should be reviewed and, in the case of TT, the choice of hints offers students support for the task.

Flexibility and challenge. The availability of tasks or sequences of tasks with various structures enables them to be organized by increasing levels of difficulty and, hence, to make them challenging for students. Each student can study at his/her own pace and achieve his/her own level of proficiency. This gradual increase in challenge seems to be important for learners' engagement with the material and can support their enquiries at several levels. Students may start with verification tasks where the question, method and answer are known and gradually move towards more open tasks, learning to find an answer, a method or even to pose a question (Banchi & Bell, 2008). In each case, they receive immediate and individualized feedback by interacting with the e-tool, which can motivate the student to continue to learn.

Support of meaningful exploration of mathematics and mathematical thinking. Both in the case of IT and of RA, the student is free to change some parameters and observe the result of this change – for example, by varying the slope of the axis or the position of the point A in Figure 2.3 or the intervals for the polar coordinates in Figure 2.4. These elements are examples of *border objects* (Moreno-Armella & Hegedus, 2009) – that is, dynamic elements that can be meaningfully explored in a digital environment. The co-action – that is, a cycle of actions and reactions between the student and the digital environment – allows the student to experience new modes of exploring mathematics and mathematical thinking. Furthermore, digital environments can have unlimited possibilities of exploration of the topic for the learners. Students can share and discuss their findings with others. This gives extra practice, and answers could be immediately verified by using online resources. It is worthwhile to note that the trial-and-error approach in designed resources is avoided by some reflection phases inserted in the educational experiment in order to encourage metacognition and reconstruction of the interaction in the online environment by the students. Moreover, the open-ended questions do not allow the trial-and-error approach.

The "playing" metaphor for accessing mathematics. Many students, in their reflection about the activities, highlighted the usefulness of "playing" with the online activities for the learning. The experience acquired through play with interactive objects prepared students for reading or listening to further explanations, moving them towards deepening further mathematical issues which now they feel are meaningful.

Multimodality and communication. Online tasks combine various modes of representation and communication, making use of various semiotic resources, not only written text but also speech, video, sound, music, etc. Such variety helps to realize interaction and meaning in a community. For instance, in the case of IT, the amount of written text is minimized and complemented by videos in order to give students the main directions and ideas that they could exemplify by interacting with diagrams.

Blended didactic organization. It is worth pointing out that the didactic organization should concern not only the f2f lectures but also the delivery of online resources. This seems obviously true if we refer to the latter as support for f2f lessons. Indeed, it will be important to deliver online resources at the right time according to when the topics at stake are addressed in the lessons. Even when the online resources are supposed to replace f2f lectures, there could be the need for a brief mathematical introduction to the topic during a f2f lesson to help the students to start working online. This has been the case with IT. Furthermore, it seems appropriate to set a sufficiently long time for the student to work on the online task. For instance, students are given one week to study the IT and work on tasks. This on the one hand gives students plenty of time to experiment with the material but on the other hand sets a deadline to complete the minimum task.

Conclusions

Blending of online resources with a f2f course may lead to a "dramatic structural shift in a student-centred direction", which "would positively contribute to students' self-efficacy" and encourage a deep-level of learning (Lahdenperä, Postareff, & Rämö, 2019). In our experience discussed in this chapter, the desired outcome is better supported by the following elements reinforced via online tasks: (1) individualization of learning, (2) alternation of immediate and delayed feedback, (3) provision of self-assessment, (4) gradual increase of challenge, (5) possibility of experimentation and exploration, (6) aspects of "play", (7) multimodality of presentation and (8) didactic organization of blending.

In view of the tetrahedron model for didactic systems in e-learning, we can describe the phenomenon of blending as follows. First, the design of online activities in M by A for S could be considered in the A-S-M face. Specifically, A considers the difficulties of S in M and exploits the mediatory role of technology inside the tetrahedron; in fact, A uses technology with the explicit didactic purpose of supporting S in overcoming his/her difficulties in M by carefully selecting e-tools and semiotic mediators and designing hints and feedback as scaffolding elements. At this point, no help from T is expected, as students are supposed to work autonomously online. However, in many cases, the technology itself can play the role of T, and/or students can consult each other as they learn the subject together and reflect on the milieu created during this work, supported by the online activities.

Second, a typical f2f setting (e.g., in lectures) could be considered in the T-S-M face. Here a course instructor in the role of T creates learning situations for S to study M, and at this point, T does not draw on any online technology but rather uses "human" teaching methods, such as gestures, metaphors and blackboard demonstrations, to help students understanding the topic.

Third, and finally, the blending of the online and f2f approaches is better viewed from the A-T-M and A-T-S faces. In the former (A-T-M), the collective of authors A and instructor T (possibly a part of A) coordinate their views about what important mathematics needs to be learnt and analyze epistemological issues. In the latter (A-T-S), A and T analyze difficulties related to the study and thus touch upon cognitive matters. Consequently, A and T coordinate their didactic approaches in teaching M to S: this could happen in many ways – for example, when the instructor T includes online elements in f2f lecture in order to demonstrate certain ideas or when T gives feedback to A, proposing to modify some online activities for students' autonomous use after the f2f lecture. In order to foster metacognition, S could be asked to reflect on his/her own relationship with the mathematical content M but also on the role played by the technology as an element of scaffolding his/her interaction with the tasks – that is, technology used as a non-human tutor. Consequently, A, possibly helped by T, uses students' feedback to modify or redesign the activities for S within the digital environment. This illustrates a dynamic character of the development of online tasks within the blending phenomenon. Among the challenges of this approach, we note that the development of online tasks possessing features (1)–(8) is a demanding and time-consuming task for the Author and the Tutor.

We conclude by suggesting some further directions for research on the design of online tasks that aim to support university students' autonomous learning of mathematics whilst having enough flexibility to be incorporated into f2f instruction if necessary: (a) inclusion of elements of scaffolding that gradually fade away, which can be realized by gradually delegating high-level functions from the computer system to the learner; (b) inclusion of elements of meta-scaffolding (Pea, 2004) that support students in selecting suitable hints in relation to their own difficulties; (c) support of the development by students of an autonomy and self-efficacy sense; (d) ways to integrate the activities of the automatic tutor and human tutor; and (e) focus on cognition and metacognition.

References

Albano, G. (2011). Mathematics education: Teaching and learning opportunities in blended learning. In A. Juan, A. Huertas, S. Trenholm, & C. Steegmann (Eds.), *Teaching Mathematics Online: Emergent Technologies and Methodologies* (pp. 60–89). Hershey, PA: Information Science Reference.

Albano, G. (2017). E-mathematics engineering for effective learning. In G. Aldon, F. Hitt, L. Bazzini, & U. Gellert (Eds.), *Mathematics and Technology: Advances in Mathematics Education* (pp. 349–370). Cham: Springer.

Albano, G., & Ferrari, P.L. (2008). Integrating technology and research in mathematics education: The case of e-learning. In F.J. Garcia Peñalvo (Ed.), *Advances in E-Learning: Experiences and Methodologies* (pp. 132–148). Hershey, PA: IGI Global.

Albano, G., Faggiano, E., & Rossi, P.G. (2018). *A didactical tetrahedron supporting codisciplinary design, development and analysis of mathematical e-learning situations*. In H.G. Weigand, A. Clark-Wilson, A. Donevska-Todorova, E. Faggiano, N. Grønbæk, & J. Trgalova (Eds.), *Proceedings of the Fifth ERME TOPIC CONFERENCE (ETC 5) on Mathematics Education in the Digital Age (MEDA)*, University of Copenhagen, Denmark, pp. 11–18.

Alessio, F., Demeio, L., & Telloni, A.I. (2019). A formative path in tertiary education through GeoGebra supporting the students' learning assessment and awareness. *International Journal for Technology in Mathematics Education*, 26(4), 191–203.

Artigue, M., & Perrin-Glorian, M.-J. (1991). Didactic engineering, research and development tool: Some theoretical problems linked with this duality. *For the Learning of Mathematics*, 11(1), 13–18.

Arzarello, F., Drijvers, P., & Thomas, M. (2012). *How representation and communication infrastructures can enhance mathematics teacher training*. Paper presented at *12th International Congress on Mathematical Education COEX*, Seoul, Korea.

Baldacci, M. (2006). *Personalizzazione o individualizzazione?* Trento: Erickson.

Banchi, H., & Bell, R. (2008). The many levels of inquiry. *Science and Children*, 46(2), 26–29.

Bardelle, C., & Di Martino, P. (2012). E-learning in secondary–tertiary transition in mathematics: for what purpose? *ZDM Mathematics Education*, 44 (6), 787–800.

Black, P., & Wiliam, G. (2009). Developing the theory of formative assessment. *Educational Assessment, Evaluation and Accountability*, 21(1), 5–31.

Cai, J. (Ed.) (2017). *Compendium for Research in Mathematics Education*. Reston, VA: National Council of Teachers of Mathematics.

Chekour, A. (2018). Computer assisted math instruction: A case study for MyMathLab learning system. In J. Silverman, & V. Hoyos (Eds.), *Distance Learning, E-Learning and Blended Learning in Mathematics Education* (pp. 49–68). Cham: Springer.

Clements, D.H., & Sarama, J. (2004). Learning trajectories in mathematics education. *Mathematical Thinking and Learning*, 6(2), 81–89.

Cusi, A., & Telloni, A. I. (2019). *The role of formative assessment in fostering individualized teaching at university level. Eleventh Congress of the European Society for Research in Mathematics Education*, Utrecht, Netherlands, pp. 4126–4129.

Cusi, A., Morselli, F., & Sabena, C. (2017). Promoting formative assessment in a connected classroom environment: Design and implementation of digital resources. *ZDM Mathematics Education*, 49(5), 755–767.

De Guzman, M., Hodgson, B.R., Robert, A., & Villani, V. (1998). *Difficulties in the passage from Secondary to Tertiary Education*. In G. Fischer & U. Rehmann (Eds.), *Proceedings of the International Congress of Mathematicians*, Bielefeld: DeutscherMathematiker-Vereinigung, III, pp. 747–762.

Díaz, C., & de la Fuente, E. I. (2006). *Assessing psychology students' difficulties with conditional probability and Bayesian reasoning.* In A. Rossmany, B. Chance (Eds.), *Proceedings of Seventh International Conference on Teaching of Statistics*, Salvador de Bahia: International Association for Statistical Education. CD ROM.

Duval, R. (2006). The cognitive analysis of problems of comprehension in the learning of mathematics. *Educational Studies in Mathematics*, 61(1), 103–131.

Geiger, V. (2006). *More than tools: Mathematically enabled technology as partner and collaborator.* In C. Hoyles, J.-B. Lagrange, L.H. Son & N. Sinclair (Eds.), *Proceedings of the Seventeenth Study Conference of the ICMI*, Hanoi Institute of Technology and DididemUniversité Paris, 7.

Hattie, J., & Timperley, H. (2007). The power of feedback. *Review of Educational Research*, 77(1), 81–112.

Kashefi, H., Ismail, Z., &Yusof, Y.M. (2012). Overcoming students' obstacles in multivariable calculus through blended learning: A mathematical thinking approach. *Procedia – Social and Behavioral Sciences*, 1, 77–86.

Kondratieva, M. (2013). Geometrical constructions in dynamic and interactive learning environment. *Mevlana International Journal of Education. Special Issue: Dynamic and Interactive Learning Environment*, 3(3), 50–63.

Kondratieva, M. (2018) *An interactive text in precalculus: Students' response.* In H.G. Weigand, A. Clark-Wilson, A. Donevska-Todorova, E. Faggiano, N. Grønbæk, & J. Trgalova (Eds.), *Proceedings of the Fifth ERME TOPIC CONFERENCE (ETC 5) on Mathematics Education in the Digital Age (MEDA)*, University of Copenhagen, Denmark, pp. 155–162

Kondratieva, M. F., & Radu, O. G. (2009). Fostering connections between the verbal, algebraic, and geometric representations of basic planar curves for students' success in the study of mathematics. *The Montana Mathematics Enthusiast*, 6(1–2), 213–238.

Lahdenperä, J., Postareff, L., & Rämö, J. (2019). Supporting quality of learning in university mathematics: A comparison of two instructional designs. *International Journal of Research in Undergraduate Mathematics Education*, 5(1), 75–96. https://doi.org/10.1007/s40753-018-0080-y

Moreno-Armella, L., & Hegedus, S. J. (2009). Co-action with digital technologies. *ZDM Mathematics Education*, 41(4), 505–519.

Mundt, F., & Hartmann, M. (2018). The blended learning concept e:t:p:M@Math: Practical insights and research findings. In J. Silverman, & V. Hoyos (Eds.), *Distance Learning, E-Learning and Blended Learning in Mathematics Education* (pp. 11–28). Cham: Springer.

Naftaliev, E., & Yerushalmy, M. (2017). Engagement with interactive diagrams: the role played by resources and constraints. In A. Leung, & A. Baccaglini-Frank (Eds.), *Digital Technologies in Designing Mathematics Education Tasks: Potential and Pitfalls* (pp. 153–174). Cham: Springer.

Nardi, E. (2017). *From advanced mathematical thinking to university mathematics education: A story of emancipation and enrichment.* In T. Dooley & G. Gueudet (Eds.), *Proceedings of the 10th Congress of the European Society for Research in Mathematics Education (CERME10)*, Dublin: Institute of Education, Dublin City University and ERME, 9–30.

Pea, R. D. (2004). The social and technological dimensions of scaffolding and related theoretical concepts for learning, education, and human activity. *Journal of the Learning Sciences*, 13(3), 423–451.

Shute, V.J. (2008). Focus on formative feedback. *Review of Educational Research*, 78, 153–189.

Simon, M. A. (1995). Reconstructing mathematics pedagogy from a constructivist perspective. *Journal for Research in Mathematics Education*, 26, 114–145.

3

QUALITY OF TASK DESIGN IN TECHNOLOGY-ENHANCED RESOURCES FOR TEACHING AND LEARNING MATHEMATICS

Ana Donevska-Todorova,

GOETHE-UNIVERSITÄT FRANKFURT AM MAIN, GERMANY

Jana Trgalová,

UNIVERSITÉ CLAUDE BERNARD LYON 1, FRANCE

Christof Schreiber and

JUSTUS-LIEBIG-UNIVERSITÄT GIESSEN, GERMANY

Teresa Rojano

CENTER FOR RESEARCH AND ADVANCED STUDIES, MEXICO

Motivation and relevance of the topic

While euphoria about technology and news headlines such as "Schools in the Post-digital Age" occupy our daily lives, research testifying the actual benefits of their implementation (before and in the digital era) in mathematics education is highly necessary. Teachers' professional activity is deeply affected by the profusion of resources available on the web: they do not need to rely only on traditional textbooks to plan their lessons but can browse the web searching for inspiration or ready-made classroom activities; they can share experience and practice with other teachers also outside of their school. These opportunities are, however, not without the challenge of securing the relevance and usability of the found resources with respect to the teachers' or curricula goals. Further questions that are of interest are as follows. How do the formulations of mathematical tasks and representations of mathematical concepts vary in the designs of different technology-enhanced resources? Do multi-touch devices, which promote gestures, have to involve audio effects, or would this increase the cognitive load of the learner? How to synchronize and balance adequate amounts of diverse dynamic representations? Yet, in the frame of this chapter, the quality of educational resources, in particular of task design, remains in focus and is a topical issue that we directly address.

This chapter firstly intends to provide an overview of theoretical and methodological approaches for investigating and reviewing the quality of technology-enhanced resources for teaching and learning mathematics. In the subsequent sections, we present and elaborate on the values and limitations of four specific kinds of technology-based task designs at three levels of education in different settings and contexts: mathematical applications and audio-podcasts for primary schools in Germany, dynamic geometry (DG) environments for secondary schools in several European countries and spreadsheets for upper secondary schools in an Anglo-Mexican context. Further, based on the analysis of these four project cases, it proposes an operational definition and criteria for the quality of task designs in the discussed four types of technology-enhanced resources that might be useful for both designers and users of those types of resources.

Theoretical background and literature review

In this section, we address the issue of quality before presenting research studies in which the quality of educational resources is tackled and specifying what we mean by the quality of technology-enhanced task design in mathematics education.

First of all, we specify what we mean by the term "resource" because there is no commonly accepted definition of the term "resource", but literature offers various conceptualizations. For example, Adler (2000) uses the term "resource" beyond its common sense as the material object and suggests thinking about it as "the verb *re-source*, to source again or differently" (p. 207). Likewise, Gueudet and Trouche (2009) consider resources in a broad sense as "everything likely to intervene in teachers' documentation work: discussions between teachers, orally or online; students' worksheets, etc." (p. 200). Digital resources are of foremost interest in the digital era. During these last years, there has been a significant evolution in considering technologies "within a range of resources available for students, teachers, teacher trainers etc." (Gueudet et al. 2010, p. 1046). These include "software, computers, interactive whiteboards, online class activities, but also more traditional geometry tools, textbooks, etc." (ibid.). Clearly, addressing the issue of the quality of a software, of a textbook or of a class activity would require different approaches and methods. In order to reduce the complexity of our endeavour in dealing with the quality of resources, we will limit ourselves to consider mathematical tasks as resources and tackle the issue of the quality of a mathematical task design in a technology-based environment.

The concept of quality in education: Current state of research

According to Mar (2013), the very concept of quality is difficult to grasp as there is little agreement on its definition. She highlights seven different definitions: a quality process or product (i) "is fit for its purpose"; (ii) "conforms to requirements"; (iii) "costs more to produce"; (iv) "is the price consumers are willing to pay for" it; (v) complies "to best known standards, processes and specifications"; (vi) "offers value for price"; and (vii) "satisfies experience". These definitions look at quality from quality assurance, marketing, manufacturing and economic points of view. In education, the issue of quality is mainly addressed in the context of e-learning; in this case, resources are online distance teaching materials. These studies usually adopt a learner-centred perspective (Ehlers, 2004; Ellis, Ginns, & Piggott, 2009). Quality indicators from this point of view refer to learner's satisfaction with the e-learning content and her/his learning (which echoes the previous definition (vii)) based on indicators such as formal aspects, contents, relation between learning objectives and contents, teaching and learning process, evaluation aspects or technical aspects (Montalvo, 2005).

Other studies investigate the quality of resources within open learning object repositories, some of which refer to the definition (ii) of quality. This is the case of the Joining Educational Mathematics repository whose quality criteria include "educational quality of content", "presentation of the content", "technological quality" in terms of usability and portability of resources, "metadata accuracy" and "availability of case studies" providing feedback from users (Caprotti & Seppälä, 2007). Others – e.g., MERLOT[1] – suggest three criteria: content quality, potential effectiveness as a teaching tool and ease of use for both students and faculty.

To sum up, the issue of evaluating resource quality is a complex one, and there is no general model describing the quality of a resource. Criteria suggested by various researchers are rather ad hoc, depending on the nature of resources (e.g., online learning courses, resources for educators), as well as on the purpose of resource quality evaluation (e.g., measuring students' satisfaction with learning material or providing authors of resources with guidelines). Nevertheless, there seems to be agreement on some criteria, such as content quality or technical quality (i.e., usability or ease of use of the resource). Yet we claim that these criteria need to be further systematized and operationalized in order to be used by resource designers in a particular subject-specific education as mathematics.

Quality of technology-enhanced task design

What do we understand under the term "quality of task design in technology-enhanced resources" for teaching and learning mathematics? On the one hand, efficient teaching and learning environments that include digital resources meet essential quality criteria for traditional ones (Krauthausen, 2012). On the other hand, Watson and Ohtani (2015) point out that theories for task design developed in digital environments can be expanded for non-digital tools. These authors state that

> any task design heuristics need to address: complementarity of feedback and mediation, relationships between pragmatic and epistemic considerations, symbiosis of mathematics and pedagogy, multiplicity of tools, and the discrepancy potential previously described. (ibid., p. 10)

Non-digital teaching and learning environments involve tasks and task formats (as collections of several tasks with similar content), and they are mutually dependent. The development and implementation of tasks, task formats and learning environments are not as straightforward as they may look from observing the visualization by Wollring (2009) but radial and two directional (Donevska-Todorova, 2018b). This means that students' activities do not necessarily have to start with a single task but rather engage with an open micro-environment. This in particular refers to environments rich in technology-based resources, as there are such that do not involve tasks a priori in their design. In these resources – for example, the tablet app Geoboard – the task design depends on teachers' creativity. This can also be noticed in a classroom implementation of a DG environment called Touchy Feely Vectors (Karnam, Agrawal, Borar, & Chandrasekharan, 2019). The design of the tasks may gain in quality when more resources are offered to the students to choose.

Rather than widening the scope with considering the quality issue in other types or resources – e.g., learning platforms for students – the quality of which has not been studied sufficiently, the chapter further sets and deepens the focus on mathematical, didactical and pedagogical criteria for the quality of specific task designs of technology-enhanced resources. The European project Intergeo (Mercat et al., 2008) investigated the quality of DG resources. An Australian project has explored the quality of existing apps by analyzing the characteristics of app designs offered by publishers (Larkin, 2013). A comprehensive model, including the review criteria for the evaluation of the quality of geometry apps for touchpad devices in the German context, has been suggested by Donevska-Todorova (2018b). Besides studies investigating quality issues of DG resources or apps, little is known about the quality of other kinds of technology-enhanced resources. For example, audio-podcasts allow the representation of mathematical content orally, while different kinds of videos use pictures and audio effects. There are, however, also videos without audio representation ("silent video tasks"; Kristinsdóttir et al., 2018) or videos using sign language (Wille & Schreiber, 2019). Furthermore, there is an important difference between whether existing video- or audio-resources are used (Peters, 2018) or created (Schreiber, 2013) by the students themselves. Further, devices such as 360-degree cameras may provide possibilities for assigning features and visualizations of geometrical objects to real-world objects and vice versa, hence enabling co-action in augmented or virtual environments. Such movements and changes of perspectives in visual observations of geometrical objects that are usually not possible with physical artefacts per se are enriched with sound and colour effects (e.g., green/red) that significantly alter the experiences students usually have with mathematics. Though these facets of apps seem to be highly relevant for the inclusion of all children in mathematics education, it would still take some time for the community to cultivate standards for future designs regarding this issue.

It is a question of whether the quality of these kinds of resources can be addressed with some of the described review criteria referring to DG resources or apps. We consider this point further in the next section of this chapter.

Research questions and research methods

The previous discussion leads us to state the following questions for a further focussed investigation.

Which criteria for the quality of task designs can be identified through the analysis of technology-based resources? Can the quality criteria specific to particular resources be synthesized/formulated and generalized in an operational definition of the quality of mathematics task design that could serve as guidelines for designers and teachers?

In our qualitative methodological approach, we have firstly considered the current state of the art related to the concept of quality in general. Then we have specified it to an educational context by referring to research on the quality of learning objects (e.g., in relation to open educational resources). These considerations have highlighted some obvious dimensions common to all research, such as content, but also missing dimensions that are important to issues in mathematics education, such as feedback or technology added value. Finally, we take a deeper look at four specific case studies/projects that have enlightened our findings.

Mathematics task design in different digital media settings and context

This section presents four different studies that consider the quality of technology-based mathematical tasks in four particular contexts: apps and podcasts for primary schools in Germany, DG tasks for secondary schools in several European countries and spreadsheet tasks for upper-secondary schools in the Anglo-Mexican context. Each case begins with a short and precise tabular overview and proceeds with a more detailed description.

Case 1: Tablet apps for primary school

Characteristics of the case/project study:	Quality criteria for reviewing tablet apps for elementary geometry in primary school
University, country:	Humboldt-Universität zu Berlin, Germany
Aim:	Developing criteria for analyzing apps by teachers
Educational level:	Primary education
Resource:	Mathematical applications for touch devices

Key elements of a "good" resource: (1) mathematical contents and relation to curriculum; (2) communication, collaboration and cooperation; (3) differentiation; (4) feedback and assessment; (5) connections and networking; and (6) logistics.

Based on the grounded methodological approach, Donevska-Todorova has framed criteria for examining the quality of digital learning environments that include tasks and task formats for tablet applications in a theoretical model with six main categories: (1) mathematical contents and relation to curriculum; (2) communication, collaboration and cooperation; (3) differentiation; (4) feedback and assessment; (5) connections and networking; and (6) logistics (Donevska-Todorova, 2018b). For example, the first category is illustrated by criteria as mathematical content presentation (clear, correct, structured, logical – e.g., geometry and competences according to national standards and curricula for primary school); task type – e.g., multiple-solution tasks (Donevska-Todorova, 2016); tasks with synchronized multiple dynamic representations as geometric, algebraic or formal-structural (Donevska-Todorova, 2018a); tasks allowing explorations; and appropriate integration of visual, audial and haptic dynamic affordances (Donevska-Todorova, 2018b). Following this manner, criteria in the other five categories are identified. Overall, this theoretical framing grounds tablet apps environments for geometry, but it is likely to expect that similar key characteristics may be relevant for akin resources as tablet apps for arithmetic or even touchpad applications in other STEM disciplines.

Task designs offered in tablet app–rich learning environments for elementary geometry are of a diverse sort. They vary from yes/no or true/false questions, through fill in the gap by choosing from suggested answers or recognizing correct figures according to given descriptions, matching pairs (e.g., plane geometric figures with their names) or entering a single number in a field (e.g., the number of sides or edges of solids up to drawing figures or exploring properties of figures). Further, plenty of apps are designed to 'push' the user in collecting points, stars or medals, which practically may only increase extrinsic motivation. While drill-and-practice exercises, common in apps about arithmetic, are limited in their didactical potential, explorative geometry tasks seem to offer diversity and meet most of the criteria categorized in the theoretical model discussed earlier (and the previous table). For example, a design of open tasks that allow possibilities for integration of digital technology (e.g., the app Geoboard), a non-digital haptic resource (e.g., Geoboard) and paper-pencil materials are as follows.

Task 1: Stretch all possible non-congruent triangles in the (3 x 3 field) of the touchpad app Geoboard, the haptic wooden material, or Geopaper (paper-pencil environment).

Task 2: Stretch all possible non-congruent squares in the (5 x 5 field) of the touchpad app Geoboard, the haptic wooden material, or Geopaper (paper-pencil environment).

The variation in the formulation of these two tasks (e.g., "stretch or sketch") depends on the technology used. In this sense of integration of diverse resources, a duo (Soury-Lavergne, 2016) of digital and non-digital artefacts are mutually complementary. Each of these tasks is rich in mathematical content and has eight possible solutions. Their openness provides possibilities for the differentiation of students' engagement through a variety of activities and solving paths through gestures and embodiment depending on the resource: app Geoboard, tactile geoboard, or Geopaper.

Case 2: Audio-podcasts for primary school

Characteristics of the case/project study:	PriMaPodcast – use of audio-podcast for mathematical education in primary school
University, country:	Justus-Liebig-University Giessen, Germany
Aim:	Developing mathematical concepts and language competences by students
Educational level:	Primary education
Resource:	Audio-recorder, Internet-resources, blog

Key elements of "good" praxis: (1) mathematical contents and relation to curriculum; (2) communication, collaboration and cooperation; (3) differentiation; (4) feedback and reflection; (5) development of concepts; and (6) mathematical creativity.

Audio-podcasts primarily focus on the aspect of orality since the process of creating an audio-podcast ends in an oral product. Schreiber (2013) developed PriMaPodcasts to serve as a tool in which oral and written forms of representation are alternated and interwoven permanently. The students undergo a process in which they develop specific mathematical content in stages up to a point where a version of the explanation is ready to be published on the Internet. The process will be explained next, describing each production step. Firstly, three aims of the mathematical audio-podcasts will be mentioned:

- Learning: The exclusive use of oral representations can support learning. This can be brought forth when learners have to deal with one specific topic in depth and are required to refrain from using any written or graphical representations, thus solely relying on oral means of communication.
- Diagnosis: A teacher gains insight into a learner's mathematical concept and understanding by listening to and analyzing a learner's spontaneous and planned recordings.
- Research: As mathematics is a written and graphically based science, focussing on oral representations is an interesting area for research. Therefore, using the recordings, the aim is to investigate to what extent learners are able to use mathematical language to express their concepts.

In step 1 ("Unexpected Recording"), a small group of students is unexpectedly confronted with a mathematical question, such as, "What is symmetry?" The students record their responses with a voice recorder.

In step 2 ("Script I"), the students research their topic and create a script. They are free to decide the format, structure and amount of detail they wish to include. In order to gather information, they are given different resources, such as the Internet, textbooks and worksheets. They also have the possibility of using concrete material.

In step 3 ("Podcast – First Version"), oral representation is essential, as the students are required to read the assigned parts of the script aloud for the recording of the first podcast version. Up to this stage, the team works independently without any intervention from the instructor.

In step 4 ("Editorial Meeting"), each group of students receives feedback from their peers and the instructor. They use this to reflect on their work in terms of content, style and language use. At this point, the instructor poses specific script-related questions and asks for more precise explanations.

In step 5 ("Script II"), the students start to improve their first script. The students have the opportunity to use the same material and artefacts again. They may discuss and share their knowledge before writing down a second script. Thus oral communication serves as a basis for written representation.

In step 6 ("PriMaPodcast"), they record a final podcast version based on the second script. This will be published on the Internet. Again, written representation serves as a basis for the oral form of representation.

The use of artefacts and various representations in the mathematics classroom is of high relevance. The aim is to produce a learning artefact, a final product of oral nature. Publishing the final version on an online blog serves as an exceptional motivation for the students. In the process, they undergo various stages in which they can expand their knowledge through discussions and learn different approaches to present them. Mathematical concepts and language competences can be strengthened.

Case 3: DG tasks for secondary school

Characteristics of the case/project study:	Intergeo European Project
Universities, countries:	Germany, Czech Rep., France, Luxemburg, Spain
Aim:	Elaboration of a quality assessment process towards the improvement of DG resources
Educational level:	Secondary education
Resource:	Mathematical tasks within a DG environment with an explicit educational goal and possibly documentation for teachers

Key elements of "good" praxis: (1) mathematical content and compliance with curriculum, (2) instrumental dimension, (3) technology added value, (4) didactical and (5) pedagogical implementation.

This study was carried out within the Intergeo European Project. One of the project aims was to elaborate a quality assessment process aiming at a continuous improvement of DG resources uploaded to the Intergeo platform by its users (Trgalova et al., 2013). Most of the resources were mathematical tasks involving DG (e.g., Geogebra, Cabri, CaRMetal, Cinderella), with an explicit educational goal and possibly with documentation aimed at teachers. To reach this goal, it was necessary to define the criteria of the quality of such resources. Around 60 criteria were formulated, grouped in nine dimensions: metadata, technical dimension, mathematical content, instrumental dimension, DG added value, didactical implementation, pedagogical implementation, integration in the teaching process and ergonomic dimension. In the sequel, we elaborate more on the dimensions that are related to the quality of the DG tasks.

Mathematical content. Criteria related to this dimension are meant to ensure not only mathematical validity of the task content, which is in line with the quality studies, but also in accordance with institutional expectations (i.e., compliance with current mathematical curricula; Chevallard, 1992).

Instrumental dimension. Research focussing on students' learning with technology highlights that the way a student uses the affordances of the technology and faces its constraints shapes her or his conceptualization of the objects she or he manipulates (Artigue, 2002). Students build technological skills (how to use the tool) in close relation to the mathematical ones (how to solve the mathematical task). The two aspects of the activity thus shape each other. Criteria related to the instrumental dimension of a DG resource aim at ensuring that the dynamic figure does produce phenomena in contradiction with the mathematical theory or the learning goal.

For example, Figure 3.1 shows a dynamic parallelogram ABCD. Measures of segment lengths and angles are set up to be rounded to the nearest integer. While dragging the vertices of ABCD, one can obtain a case where AB and BC have seemingly the same length (Figure 3.1); ABCD is therefore a rhombus. However, its diagonals do not appear perpendicular, which contradicts the mathematical theory. The choice of displaying the measurements of lengths and angles rounded to the nearest integer explains this contradiction, as, in fact, AB can measure 5.3 units, while the measure of BC can be 4.9 units.

This example shows that it is necessary to verify that the dynamic figure (with associated settings) does not convey misconceptions, which led to suggest the following criterion: "Numerical values (lengths and angles measures) do not hinder mathematical activity". It is also important to check the coherence between the proposed task, intended learning and associated dynamic figure, which are expressed by the following criterion: "The figure behaves consistently with the mathematical task".

DG added value. Criteria related to this dimension aim at evaluating how DG potentialities are exploited, and to what extent DG contributes to the targeted learning compared to traditional paper-pencil tasks. The main assumption is that technology-enhanced tasks have to be designed in order to take advantage of technology (Laborde, 2008). The criterion "the activity cannot be transposed to paper and pencil" points out this aspect. The following criterion addresses specific features of DG, offering an added value to the resource: "DG helps the user explore, experiment and conjecture". Other criteria focus on the role of the drag mode, drawing on research studies that identified different purposes of the drag mode leading to

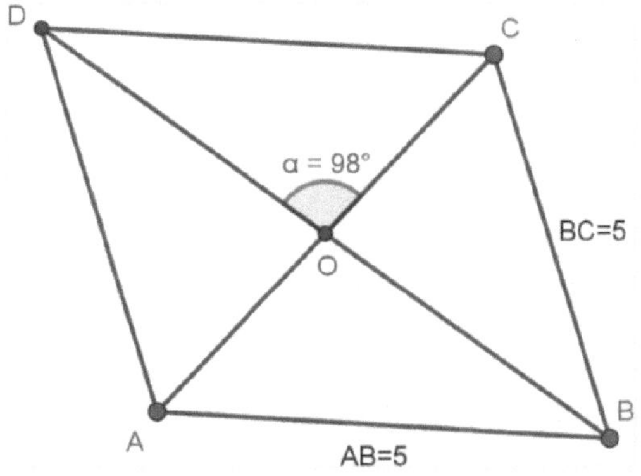

FIGURE 3.1 DG drawing displaying mathematically incorrect properties (created with GeoGebra: geogebra.org)

different mathematical conceptualizations (e.g., Restrepo, 2010). For instance, the criterion "dragging is used to explore validity conditions of a theorem" refers to this specific use of the drag mode in soft constructions (Healy, 2000).

Didactical implementation of the task. Criteria associated to this dimension are underpinned mainly by the theory of didactical situations (Brousseau 1997) offering tools for analyzing students' activities and teachers' roles and by the instrumental approach (Rabardel 2002) providing a framework for students' instrumented activity analysis. The notion of feedback is crucial here, and it is addressed by criteria such as "feedback provided by the software helps students in the problem solving activity". Likewise, the notions of devolution and institutionalization (Brousseau 1997) are behind the criteria "hints are given on ways to make students start solving the activity" and "advice is given to determine how and when to synthesize findings". The latter are related to the documentation aimed at teachers rather than the task itself.

Pedagogical implementation of the task. Trouche (2004) stresses the essential role of the teacher in a successful integration of technology and introduces the concept of instrumental orchestration for analyzing the teacher's management of activities in a technology-rich environment. Drawing on the orchestration perspective, a quality resource should provide the teacher with information related to the classroom orchestration conducive to the intended learning, leading to the following criterion: "Possible configurations (one computer per student or projector…) are described".

Case 4: Algebra and modelling spreadsheets for primary, secondary, tertiary and university levels

Characteristics of the case/project study:	Anglo-Mexican Spreadsheets Algebra Project and the role of spreadsheets within school-based mathematical practices in sciences.
Universities, countries:	University of London and Bristol University (United Kingdom); Center for Research and Advanced Studies (Cinvestav–Mexico).
Aim:	The use of spreadsheets for the teaching and learning of algebra and parameterized modelling
Educational level:	Primary, tertiary and university levels
Resource:	Algebra and modelling activities designed with research ends and for implementation in the classroom

Key elements of "good" praxis: (1) spreadsheet activities designed for in-school algebra and modelling teaching and learning, (2) didactical design to help primary and secondary students in the transition from intuitive and numerical methods towards symbolic and algebraic methods and (3) didactical design to foster tertiary and university students' mathematical modelling competencies based on parameterized versions of models of phenomena of different science areas.

Outcomes from the Anglo-Mexican Spreadsheets Algebra Project show that primary and secondary students develop algebraic ideas and notions with a numerical approach in that environment incorporating the use of formulas. One of the studies carried out in the framework of this collaborative project was done simultaneously with two groups of pre-algebraic students (10- to 11-year-olds). One group was in Mexico and the other in the United Kingdom. The students enrolled in a sequence of spreadsheet activities focussed on the ideas of function and inverse function, as well as on the solution of algebra story problems (Rojano & Sutherland, 1994). It is worth noting that despite the differences in the mathematics teaching approaches in the two countries, there were no notorious differences in the performance of both groups of students. The latter may be attributed to the general task design principles applied: (1) to take advantage of the environment to help students systematize their intuitive numerical approaches and (2) to foster processes of transition towards more formal algebraic methods.

On the other hand, in the Mexican-British project, The Role of Spreadsheets within the School-Based Mathematical Practices, research was done on the role of spreadsheets as a mathematics modelling tool with two groups of tertiary-level students aged 16 to 18, one in each country (Molyneux-Hodgson et al., 1999). The pre-experimental phase results revealed important differences in student preferences for given external representations of situations and phenomena (for instance, we

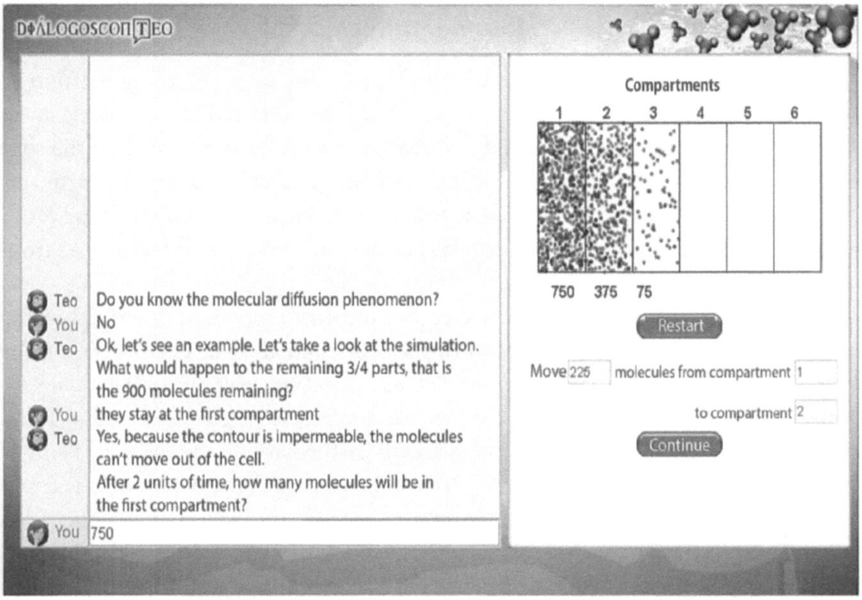

FIGURE 3.2 Screen with a simulation of the molecular diffusion in a cell (image courtesy of Laboratorio de Innovación en Tecnología Educativa, México)

observed a marked preference for the use of graphic representation amongst students in the United Kingdom and for the use of algebraic representation amongst students in Mexico). Nevertheless, despite the fact that the experimental work with spreadsheets did not significantly change those differences, at the end of the study, the students enhanced their ability to shift from one type of representation to another and recognized the value of using a diverse repertoire of representations – including graphic, algebraic and numerical representations – when working with modelling activities. These outcomes can be attributed in part to the main principle underlying task design: to take advantage of the availability of a set of representation systems dynamically hyperlinked to each other, to explore or create parameterized models of situations and phenomena without the necessity of dealing with formal mathematical models that involve advanced mathematical knowledge (Rojano & García-Campos, 2016). The modelling task simulating *molecular diffusion in a cell* (Figure 3.2) used in the aforementioned study illustrates this idea.

The phenomenon consists of considering a simplified cell (in two dimensions) with six compartments. Students could explore the diffusion phenomenon on a short and long term, using either the formulas or the numerical inputs in the spreadsheet or the corresponding graphical representation.

Results of the study and discussion

Criteria for the quality of task design in technology-enhanced resources for teaching and learning mathematics

This section summarizes the discussion of the previous cases regarding the potentials and limitations of specific types of technology-based task designs at three levels of education in different settings and contexts. As a tool for the analysis of the four cases we, "zoom in" on the edges of the "student artefact" and "teacher artefact" of a didactical tetrahedron (Donevska-Todorova & Trgalova, 2017). For an easier overview, Table 3.1 synthesizes the criteria considered in each of the four cases from students' and teachers' perspectives.

Regardless of context, whether in German primary schools, as described in the first two cases, or in a European secondary school context, as in Case 3, the *validity of the mathematical content presented in relation and accordance with corresponding curricula* shows to be an important criterion for the task design with digital tools. Cases 1 and 4 show task designs with a particularly selected mathematical content (i.e., elementary geometry and algebra, respectively). While the designs for podcasts in Case 2 and for spreadsheets in Case 4 consider the subject-related and didactical contents from students' perspectives, such contents in apps for touch devices in Case 1 and in DGS in Case 3 have been analyzed also from teachers' perspectives. This finding is in coherence with the theoretical considerations about the quality of educational resources presented earlier in this chapter.

Further criterion that both projects for primary school mathematics education have in common is the possibility of the digital designs to strengthen *communication*,

TABLE 3.1. Summary of criteria for the quality of task designs

Case	*Type of digital resource*	*School level/perspective/context*	*Quality criteria considered in task designs*
1	Apps	primary school/ students and teachers/ Germany	– mathematical contents and relation to curriculum (e.g., elementary geometry) – communication, collaboration and cooperation – differentiation – feedback and assessment – connections and networking – logistics
2	Audio-podcasts	primary school/ students/ Germany	– different contents related to curriculum – communication, collaboration and cooperation – differentiation – feedback and reflection – development of mathematical concepts – mathematical creativity
3	DGS	secondary school/ teachers/Europe	– content validity and compliance with curriculum – instrumental quality – DG added value – didactical implementation including learner's initiative and feedback – pedagogical implementation
4	Spreadsheets	secondary school/ students/Mexico and the United Kingdom	– activities designed for algebra and modelling – support students in different transitions (between different educational levels, or between different representations) – foster students' mathematical modelling competencies based on parameterized versions of models of phenomena of different science areas

collaboration and *cooperation* between pupils. This shows that the potential of digital tools in supporting different social and organizational forms of collaborative learning is considered in task designs for apps and podcasts through forms of games or competitions. In the case of spreadsheets or DG, the tasks can be enacted individually or in small groups of students, depending on teachers' practices or even on the number of computers available. *Differentiation* is thought to mean diversity of tasks types (e.g., open tasks or multiple-solution tasks) or plenty of levels of difficulties within one task or tasks and allowing the combining of non-digital with digital artefacts through different sensory canals (tactile, audio, video) for their solutions is another common feature in Cases 1 and 2. Sound and colour effects in apps, possibilities for co-action in augmented reality apps and oral, written and externalization of thoughts through listening, recording, scripting and blogging in podcasts are further options for differentiation. These appear to be relevant for task designs for primary school. The affordances of perceiving information through single or *synchronized multiple dynamic representations* have been considered in the review criteria for the quality of the task designs in all four cases.

The first three cases show that the *feedback* feature is commonly considered in apps, podcasts and DGS. Novel designs of apps attempt to include "intelligent agents" that may provide individual feedback and enable personalized guidance to each student. Recent developments of DGS in the form of apps that run on portable devices try to bring advantages of both digital resources together (e.g., (multiple) finger touch substituting the mouse drag). This characteristic is often connected to *assessment* (e.g., in Case 1, the reflection in Case 2 and the learner's initiative in Case 3).

Like differentiation, synchronized multiple dynamic representations, feedback and assessment also fit into the *technology added value* category, which is further analyzed through the Substitution, Augmentation, Modification, Redefinition (SAMR) model in the next section.

Operational definition for quality of task design in technology-enhanced resources for teaching and learning mathematics

This section proposes our view of the quality of mathematical technology-enhanced task design based on the four cases reported earlier. These cases are contrasted in terms of the nature of the technology involved, which to a certain extent enables the generalization of our task design quality considerations.

Not surprisingly, the *mathematical content* of the task appears as a common concern of the four cases, which is also in line with other research findings. This dimension addresses three aspects. First, the *validity of the content* (i.e., the task needs to be designed in a way that students do not risk developing misconceptions). Second, the task should *comply with the mathematics curricula*. Third, the *digital support* (e.g., associated DG or spreadsheet file) *is coherent with the educational goal*. This last criterion echoes the instrumental dimension whose importance was highlighted by the Intergeo consortium (Case 3). Indeed, the example in Figure 3.1 shows that although the task itself is mathematically sound, the DG figure displays mathematically incorrect geometric drawings.

Since our focus is on technology-enhanced mathematical tasks, questioning how technology enhances the task arises naturally, leading us to formulate a *technology added value* dimension of task quality. The four cases brought forward the potential of technology to display *interconnected multiple representations* of mathematical concepts, which helps mathematical conceptualizing. In reference to the SAMR model,[2] we suggest identifying the level at which technology is used in the task:

- *Substitution:* "Technology acts as a direct substitute with no functional change".
- *Augmentation:* "Technology acts as a direct substitute with a functional improvement".
- *Modification:* "Technology allows for significant task redesign".
- *Redefinition:* "Technology allows for the creation of new tasks, previously inconceivable".

Clearly, the higher the level (from S to R), the more the technology potential is exploited.

TABLE 3.2. Operationalization of the quality of task design in digital resources for teaching and learning mathematics

	Task dimension	*Quality criteria*
I	Mathematical content	1. The content leads to valid mathematical knowledge (i.e., students do not risk developing misconceptions) 2. Compliance with mathematics curriculum 3. Coherence between the learning goal and digital support
II	Technology added value	1. The task involves technology use at the substitution, augmentation, modification or redefinition levels (S – A – M – R) 2. Technology allows for interconnected multiple representations of mathematical concepts
III	Student's perspective	1. The technology-based task leaves enough room for students to take initiative 2. Technology provides feedback supporting students' mathematical activity 3. Technology fosters student's (a) motivation, (b) autonomy, (c) written or oral expression, (d) self-assessment and (e) collaboration, cooperation
IV	Teacher's perspective	1. The educational goal of the task is clear or can be easily perceived 2. Possible orchestration of the task is suggested 3. Possible differentiation of the task is suggested

The four cases also highlighted the importance of technology affordance supporting students' mathematical activity. This leads us to include *student's perspective* on the definition of task design quality, which is in line with other research studies. We claim that quality task design should yield mathematical tasks, leaving enough room for *student's initiative*. Open-ended or multiple-solution tasks reported in some of the cases satisfy this requirement. Student's mathematical activity needs to be supported by *feedback* from the technology-enhanced *milieu* (Brousseau, 1997). Finally, the role of technology is also to foster students' motivations (e.g., gamification features, as in the case of apps), autonomy (in relation to feedback), written or oral expressions (as in audio-podcasts), self-assessments and interactions with peers.

To facilitate appropriation of the designed tasks by the teachers, we add *teacher's perspective* as another dimension contributing to the quality of the task design. This dimension includes *explicit educational goals*, suggestions about *classroom orchestration* (pedagogical scenario) and possible *differentiation* of the task.

The task dimensions and associated criteria that constitute our definition of task design quality are summarized in Table 3.2.

Conclusion and perspectives

Recent developments of the resource approach to mathematics education (e.g., Pepin, Gueudet, & Trouche, 2017) stress the need to pay attention to existing resources and teachers' interactions with them. Considering that teachers' work with resources shapes their professional practices, the issue of resource quality is crucial. In this chapter, we attempted to contribute to research on this issue by considering technology-enhanced mathematical tasks as resources.

We have discussed quality criteria that refer to four specific types of digital resources – apps, audio-podcasts, DG and spreadsheets – addressing all educational levels, from primary to tertiary, from a theoretical and empirical viewpoint using qualitative approaches and four cases. Based on these case studies, we have proposed our view of the quality of technology-based mathematical task design summarized in four aspects of the design: mathematical content quality, technology added value and students' and teachers' perspectives, which directly address the posed research question. Although these criteria are based on specific technology, we believe that they are general enough to apply to other kinds of technology involved in mathematical tasks.

The purpose of suggesting an operational definition of task design quality was twofold. First, we believe that it can serve as guidelines for designers of digital resources, as it points out the main dimensions of a resource deemed of "good" quality. Second, our main targeted public is mathematics teachers. Indeed, the European Framework for the Digital Competence of Educators (Redecker & Punie, 2017, p. 20) claims,

> One of the key competences any educator needs to develop is to come to terms with this variety [of digital educational resources], to effectively identify

resources that best fit their learning objectives, learner group and teaching style, to structure the wealth of materials, establish connections and to modify, add on to and develop themselves digital resources to support their teaching.

We believe that having at their disposal explicit resource quality criteria may help teachers identify relevant and meaningful resources, modify them or design new resources for their mathematics teaching. Empirical studies conducted in the framework of the Intergeo Project, in relation to Case 3, highlight the strong impact of teachers' work on the evaluation of DG resources on their professional development (Trgalova & Jahn, 2013). An ongoing design-based research involving pre-service teachers in one of its phases and in-service teachers through a teacher development program in a later phase, related to Case 1, emphasize the teacher's role in selection, redesign and dissemination of criteria for efficient task designs (Donevska-Todorova, 2018b).

One may argue that the suggested definition of mathematical task design quality is rather research-based and theory-driven. Further empirical studies involving teachers are needed to highlight whether teachers' perceptions of the quality of resources are mirrored in our definition and to what extent that definition is operational and leads to improved task design. Our research is also based on four particular kinds of resources, which raises the question of the generalizability of this definition to other types of digital resources, such as e-textbooks. It is also a question of what extent these criteria apply to digital task designs in other subject matters (e.g., natural sciences or language). These considerations open avenues for further research.

Notes

1 MERLOT: https://www.merlot.org/merlot/.
2 "InstructionalDesign/SAMR/Model/WhatIstheSAMRModel," https://en.wikiversity.org/wiki/Instructional_design/SAMR_Model/What_is_the_SAMR_Model%3F.

References

Adler (2000). Conceptualising resources as a theme for teacher education. *Journal of Mathematics Teacher Education*, 3, 205–224.

Artigue, M. (2002). Learning mathematics in a CAS environment: the genesis of a reflection about instrumentation and the dialectics between technical and conceptual work, *International Journal of Computers for Mathematics Learning*, 7, 245–274.

Brousseau, G. (1997). *Theory of Didactical Situations in Mathematics* (vol. 19). Mathematics Education Library Series. Springer.

Caprotti, O., & Seppälä, M. (2007). Evaluation criteria for eContent quality. JEM – Joining Educational Mathematics ECP-2005-EDU-038208. Deliverable, D2.1.

Chevallard, Y. (1992), Concepts fondamentaux de la didactique: Perspectives apportées par une approche anthropologique. *Recherches en Didactique des Mathématiques*, 12(1), 77–111.

Donevska-Todorova, A. (2016). *Procedural and conceptual understanding in undergraduate linear algebra*. In E. Nardi, C. Winsløw, & T. Hausberger (Eds.), *Proceedings of the First Conference of the International Network for Didactic Research in University Mathematics (INDRUM 2016)*, Montpellier, France: University of Montpellier and INDRUM, 276–285. https://hal.archives-ouvertes.fr/hal-01337932.

Donevska-Todorova, A. (2018a). Fostering students' competencies in linear algebra with digital resources. In S. Stewart et al. (Eds), *Challenges and Strategies in Teaching Linear Algebra* (pp. 261–276). Cham: Springer.

Donevska-Todorova, A. (2018b). Towards a theoretical foundation for quality tablet app-enriched learning environments in primary school mathematics education. *International Journal for Technology in Mathematics Education*, 26(3), 121–129.

Donevska-Todorova, A., & Trgalova, J. (2017).*Learning mathematics with technology. A review of recent CERME research*. In T. Dooley, & G. Gueudet (Eds.) *Proceedings of the Tenth Congress of the European Society for Research in Mathematics Education (CERME10)*, Dublin, Ireland: DCU Institute of Education and ERME, 2539–2546.

Ehlers, U. (2004). Quality in e-learning from a learner's perspective. *European Journal of Open and Distance Learning, I.* http://www.eurodl.org/materials/contrib/2004/Online_Master_COPs.pdf).

Ellis, R. A., Ginns, P., & Piggott, L. (2009). E-learning in higher education: some key aspects and their relationship to approaches to study. *Higher Education Research & Development*, 28(3), 303–318.

Gueudet, G., & Trouche, L. (2009). Towards new documentation systems for mathematics teachers? *Educational Studies in Mathematics*, 71(3), 199–218.

Gueudet, G., Bottino, R.M., Chiappini, G., Hegedus, S., & Weigand, H.-G. (2010). *Introduction. Technologies and resources in mathematical education*. In V. Durand-Guerrier et al. (Eds.), *Proceedings of the Sixth Congress of the European Society for Research in Mathematics Education (CERME 6)*, Lyon, France, INRP, 1046–1049.

Healy, L. (2000). *Identifying and explaining geometrical relationship: Interactions with robust and soft Cabri constructions*. In T. Nakahara & M. Koyama (Eds.), *Proceedings of the 24th PME Conference*, Hiroshima, Japan, 103–117.

Karnam, D., Agrawal, H., Borar, P., & Chandrasekharan, S. (2019). *The Affordable Touchy Feely Classroom: Textbooks embedded with Manipulable Vectors and Lesson Plans augment imagination, extend teaching-learning practices*. In Lund, K. et al. (Eds.), *Proceedings of 13th International Conference on Computer Supported Collaborative Learning*, Lyon, France: International Society of the Learning Sciences, 1, 488–495.

Krauthausen, G. (2012). *Digitale Medien im Mathematikunterricht der Grundschule*. Berlin: Spektrum, Springer.

Kristinsdóttir, B., Hreinsdóttir, F., & Zsolt, L. (2018). *Realizing students' ability to use technology with silent video tasks*. In A. Clark-Wilson et al. (Eds.), *Proceedings of the Fifth ERME Topic Conference on Mathematics Education in the Digital Age (MEDA)*, Copenhagen, Denmark: University of Copenhagen, 163–170.

Laborde, C. (2008). Multiple dimensions involved in the design of tasks taking full advantage of dynamic interactive geometry. In A. P. Canavarro et al. (Eds.), *Tecnologias e Educação Matemática* (pp. 29–43). Lisboa: SEM/SPCE.

Larkin, K. (2013). *Mathematics education. Is there an app for that?* In V. Steinle, L. Ball, & C. Bardini (Eds.), *Mathematics Education: Yesterday, Today and Tomorrow (Proceedings of the 36th Annual Conference of the Mathematics Education Research Group of Australasia)*, Melbourne, VIC: MERGA, 426–433.

Mar, A. (2013). 7 definitions of quality. *Simplicable*. Online https://business.simplicable.com/business/new/7-definitions-of-quality.

Mercat C., Soury-Lavergne S., and Trgalova J. (2008). Quality assessment plan. *Intergeo, Deliverable N* D6, 1.

Molyneux-Hodgson, S., Rojano, T., Sutherland, R., & Ursini, S. (1999). Mathematical modeling: The interaction of culture and practice. *Educational Studies in Mathematics*, 39(1–3), 167–183.

Montalvo (2005). Conceptual model for ODL quality process and evaluation grid, criteria and indicators. e-Quality Project 110231-CP-1-2003-FRMINERVA-M. Deliverable, D2.2. http://www15.uta.fi/projects/e-quality/pdf/deliverables/e-quality_d2p2.pdf.

Pepin, B., Gueudet, G., & Trouche, L. (2017). Refining teacher design capacity: Mathematics teachers' interactions with digital curriculum resources. *ZDM Mathematics Education*, 49(5), 799–812.

Peters, F. (2018). *Auditory material for specialised language support in mathematics education*. In A. Clark-Wilson et al. (Eds.) *Proceedings of the Fifth ERME Topic Conference on Mathematics Education in the Digital Age (MEDA)*, Copenhagen, Denmark: University of Copenhagen, 195–200.

Rabardel, P. (2002). People and technology – A cognitive approach to contemporary instruments. Université Paris, 8. https://hal.archives-ouvertes.fr/file/index/docid/1020705/filename/people_and_technology.pdf.

Redecker, C. (2017). European Framework for the Digital Competence of Educators (DigCompEdu). Joint Research Centre Science for Policy Report. EU: Luxembourg Publications Office.

Restrepo, A. M. (2010). *Genèse instrumentale du déplacement en géométrie dynamique chez des élèves de 6ème*. Editions Edilivre – Collection universitaire.

Rojano, T., & García-Campos, M. (2016). Teaching mathematics with an intelligent support in natural language. Tertiary education students working with parameterized modeling activities. *Teaching Mathematics and Its Applications*, 36, 18–30. https://doi.org/10.1093/teamat/hrw009.

Rojano, T., & Sutherland, R. (1994). *Towards an algebraic notion of function: The role of spreadsheets*. In *Proceedings of the 16th Annual Meeting of the North American Chapter of the International Group for the Psychology of Mathematics Education (PME-NA)*, USA: Louisiana State University, I, 278–284.

Schreiber, C. (2013). *PriMaPodcast – Vocal representation in mathematics*. In B. Ubuz, Ç. Haser, & M. A. Mariotti (Eds.) *Proceedings of the 8th Congress of the European Society for Research in Mathematics Education (CERME 8)*, Antalya, Turkey: Middle East Technical University, 1596–1606.

Soury-Lavergne, S. (2016). *Duos of Artefacts, connecting technology and manipulatives to enhance mathematical learning*. In *Proceedings of the 13th International Congress on Mathematics Education ICME13*, Hamburg, Germany, 24–31.

Trgalova, J., & Jahn, A. P. (2013), Quality issue in the design and use of resources by mathematics teachers. *ZDM – The International Journal on Mathematics Education*, 45(7), 973–986.

Trouche, L. (2004). Managing the complexity of human/machine interactions in computerized learning environments: Guiding students' command process through instrumental orchestrations. *International Journal of Computers for Mathematical Learning*, 9, 281–307.

Watson A., & Ohtani M. (2015). Themes and issues in mathematics education concerning task design: Editorial introduction. In A. Watson, & M. Ohtani (Eds.) *Task Design in Mathematics Education* (pp. 3–15). Cham: Springer.

Wille, A., & Schreiber, C. (2019). *Explaining geometrical concepts in sign language and in spoken language – a comparison*. In U.T. Jankvist, M. Van den Heuvel-Panhuizen, & M. Veldhuis (Eds.) *Proceedings of the Eleventh Congress of the European Society for Research in Mathematics Education (CERME 11)*, Utrecht: Freudenthal Group & Freudenthal Institute, 4609–4616.

Wollring, B. (2009). Zur Kennzeichnung von Lernumgebungen für den Mathematikunterricht in der Grundschule. In A. Peter-Koop, G. Lilitakis, & B. Spindeler (Eds.) *Lernumgebungen – Ein Weg zum kompetenzorientierten Mathematik in der Grundschule* (pp. 9–23). Offenburg: Mildenberger Verlag.

4

TOWARDS PRAGMATIC THEORIES THAT UNDERPIN THE DESIGN OF TEACHER PROFESSIONAL DEVELOPMENT CONCERNING TECHNOLOGY USE IN SCHOOL MATHEMATICS

Eleonora Faggiano,

UNIVERSITY OF BARI ALDO MORO, ITALY

Helena Rocha,

UNIVERSIDADE NOVA DE LISBOA, PORTUGAL

Ana Isabel Sacristán and

CENTER FOR RESEARCH AND ADVANCED STUDIES (CINVESTAV), MEXICO

Marisol Santacruz-Rodríguez

UNIVERSIDAD DEL VALLE, COLOMBIA

Introduction

Although the complex theoretical ideas that help us to understand and explain how teachers learn to use digital mathematical tools, and the knowledge and practice that ground such use, are reasonably developed, far less is known about the effective design of teacher professional learning activity in this respect. This chapter aims to make more explicit the grounded or "pragmatic theories" that inform the design of professional learning for teachers that intends to support them to exploit technological affordances for the purpose of learning, teaching and assessing mathematics. We interpret professional learning as the in- and out-of-school activities that teachers engage in, that further promote their professional knowledge, which we consider in the frame of the Mathematical Pedagogical Technology Knowledge or MPTK model (Thomas & Hong, 2013; Clark-Wilson & Hoyles, 2018).

There are no "global" theories to support the design of professional development (PD) concerning technology use in mathematics education (Clark-Wilson & Hoyles, 2018). However, much "craft knowledge" on the part of the mathematics teacher educator/research communities has emerged, which looks promising but currently lacks connecting ideas that enable the knowledge to be shared more widely. By exploring the evolution of, and implications from mainly large-scale national PD initiatives, we aim to arrive at a more theoretical grounding for the topic, supported by case examples that exemplify the most pertinent aspects.

The chapter aims to offer teacher educators and authorities an empirically informed and research-based approach (our set of pragmatic theories) on how to prepare teachers of mathematics for present and future digitally enhanced contexts so that they can cope with emergent technologies – mobile technologies, augmented and virtual reality, the use of multiple screens, virtual collaborations, etc. – and develop workable pedagogies for their uses.

Some representative international projects

We begin by reflecting on significant country efforts in Colombia, Italy, Mexico and Portugal as a context to examine the respective designs and processes of PD for teachers of mathematics. Some projects are described next (with comparative analyses provided in Table 4.1); we consider these projects as representative because they were important in each of the countries, either because they were very large-scale, were substantial designs or had significant impacts.

For Colombia, we selected three nationwide PD programmes with varying objectives: INTC[1] (1998–2004), CIER[2] (2013–2015) and CPE[3] (2014–ongoing). INTC sought to integrate the use of digital technologies into mathematics classes in secondary schools. CIER focussed on both diversifying the creative uses of digital technologies in teaching in schools (including mathematics) and supporting the technological capabilities of each region of the country. CPE sought to provide technological tools, training and support to educational communities in an environmentally conscious manner by reusing discarded hardware from other sectors.

For Italy, we chose *m@t.abel*[4] (2005–2012), launched and supported by the Ministry of Education and developed through a joint collaboration with university researchers. This project consolidated several communities of practice (CoPs) with a blend of teacher training activity and classroom experimentation to offer infrastructural structures (an online platform and support materials) and expert tutor support. Based on the experience of m@t.abel, a further (online based) project, DiFiMa[5] (2008–ongoing), has been implemented at the University of Turin with the aim of reaching mathematics and physics schoolteachers all over the country through the use of an online platform. The DiFiMa project is intertwined with the local initiatives developed within the huge national project PLS,[6] which, since 2005, has enabled the participating universities to organize different initiatives that aim to promote the attractiveness of the scientific university degree courses. For mathematics teacher education, many of these initiatives aim to foster the integration of technology in the teaching and learning processes. DiFiMa highlights the possibilities of technology use for mathematics in schools through periodic face-to-face meetings organized by the university researchers for local teachers. In addition, an online digital platform is used by the organizing team to train the volunteer teachers who provide local tutoring support on the national scale. In particular, the PD programme offers teaching materials with guidelines and distance support and, more recently, some Massive Open Online Courses (MOOCs; Aldon et al., 2019).

For Mexico, we chose two national efforts and one local effort that took place in the early 21st century and were significant because of their scale: the Teaching Mathematics (and Science) with Technology programme for lower-secondary school (EMAT, 1998–2007)[7]; Enciclomedia (2002–2007) for primary school[8] and a local university master's PD programme. Both national projects aimed to integrate technology into schools and, therefore, included accompanying teacher training. However, their conceptions and designs were radically different: EMAT was carefully designed by researchers and planned to be gradually introduced (Sacristán & Rojano, 2009). Enciclomedia was a politically led decision to create digitized textbooks with accompanying interactive applications and hardware. The local university master's PD programme was for around 100 in-service teachers that took place from 2005 to 2009 at the request of one state's government; we call it M.Ed. (Sacristán, Sandoval, & Gil, 2011). An overview of the aforementioned Mexican projects is given in Trouche et al. (2012). We also mention very briefly, but do not analyze, a more recent project for general technology integration into Mexican schools: the Digital Abilities for All (HDT) programme (2011–2017).[9]

For Portugal, we selected three PD projects that had a notable impact on Portuguese schools: MINERVA[10] (1985–1994), *Programa de Acompanhamento* or Follow-Up of Syllabus (FUS) implementation (1998–2001) and *Plano da Matemática* (PM; 2006–2012). The MINERVA project was the first major project on technology, aiming to integrate technology in the teaching and learning of all subjects at all levels (Ponte, 1994). FUS, which followed the implementation of a new high school mathematics syllabus, intended to promote the teachers' discussions about their practices, addressing, namely, the role of the students and the use of new materials and resources where technology was an important element (Teixeira, 2004). PM was a national programme with voluntary teacher enrolment, aiming to improve students' outcomes based on a change of teachers' practices, where technology was usually an important element. The specific characteristic of PM was the recognition of the role of the teachers, leaving to them the creation of small CoPs and the decision about the specific focus of each of those communities (GTM, 2019).

Themes and aspects elicited by the projects

Our method to elicit the themes and aspects from the projects was twofold. Our process of analysis was inspired by the work of Tondeur et al. (2012), whose systematic review identified a number of key themes for content and delivery methods to prepare teachers, in general, to integrate technology into their practice. Alongside, we considered aspects related to teachers' knowledge. For instance, to analyze how teacher training is articulated to mathematics syllabi, we used categories from the Mathematical Knowledge for Teaching or MKT model (Ball, Thames, & Phelps,

2008), and, to analyze if the focus of the PD programmes promoted teachers' knowledge for technological integration in math classes, we took into account the various components of the MPTK model (Thomas & Hong, 2013; Clark-Wilson & Hoyles, 2018).

Each of the contributing authors first analyzed their country's projects separately and then compared their analysis with the classifications for the other projects. We then iterated on the emerging categories to derive our own general set of themes, aspects and sub-aspects – those we considered as most important and characteristic and that related to the PD of teachers for technology use in their mathematics teaching practice. The final aspects and sub-aspects are organized in Table 4.1 in the Appendix section.

In the descriptions that follow, we use elements of the projects in the four countries to illustrate lessons learned related to each theme and aspect (e.g., similarities and differences, barriers or opportunities, what works, what does not, what impacts, what is needed/missing) that give pointers –i.e., pragmatic suggestions/theories – for future PD implementations that would aim to develop MPTK. The aspects are related to the following five themes:

1. **Vision/goals and policies:** The vision, goals and policies that drive a PD programme, as well as to its scale (e.g., national, regional).
2. **Focus:** The focus of the PD programme in terms of whether it is specific for math, general for all school contents or technology in general and of who defines that focus.
3. **Strategies and methods:** The strategies and methods of, and for, PD programmes and/or policies that affect them.
4. **Constraints or allowances and incentives:** The factors that restrict or facilitate participation in a PD programme and/or technology integration.
5. **Evaluating the impact:** The processes to assess the impact of the PD programme (e.g., on teachers' practice) and how that informs future adaptations (e.g., proposals for improvement) or other PD programmes.

Each aspect (and sub-aspect) that was derived from the themes is now described with examples from the national experiences that helped us recognize how those aspects can be implemented for optimal results in future PD programmes. These aspects and the evidence used to determine them serve to make visible our "pragmatic theories", both for the design and for evaluating PD programmes that focus on integrating technology in mathematics education.

Vision/goals and policies of the PD programmes

Some projects have a vision at a teacher level (e.g., INTC in Colombia, DiFiMa in Italy, Mexico's M.Ed. and PM in Portugal); that is, their aim is to train teachers to either improve their mathematics teaching in general or to help them to integrate technology into their practice. Many initiatives are derived from national policies that aim to integrate technologies in general in schools (e.g., MINERVA in Portugal and Enciclomedia in Mexico). Others, either derived from national or local policies, focussed more specifically on technology integration for mathematics or science education (e.g., EMAT, in Mexico).

However, top-level initiatives that do not take into account the specific necessities of schools, nor of the different subjects (e.g., mathematics), as was the case of HDT in Mexico in the last decade, have been shown to be ineffective (Trouche et al., 2012; Macías-Herrera, 2014).

However, it seems that when top-level initiatives have been coordinated with subject matter experts (i.e., in mathematics education) and developed programmes that have had more specific goals (such as promoting changes in school cultures and teachers' practices), they have been more effective. Such were the cases of EMAT in Mexico, FUS in Portugal, INTC in Colombia and m@t.abel in Italy. In addition, these projects were "research-based" (i.e., taking into account research results in their design and implementation), expert-supported (some with connections to academic institutions) and had associated research projects. Such projects are thus more likely to attend to the needs and specificities of the specific subject area (mathematics) and of schools and teachers. In the case of EMAT, although it was a national programme, it was designed with a scaled implementation from a few schools and regions to expand to a national level over the course of several years, thus allowing for necessary research-informed iterations to the implementation. This scaled approach was also adopted for the PM in Portugal, which started as a PD programme for volunteer teachers and ended on a national scale. The INTC project in Colombia was also national, had the support of several regional universities and was associated with several research projects developed in the participating universities. This project focussed on promoting change in the practices of secondary school mathematics teachers through strengthening their professional knowledge (e.g., on the design of teaching activities) and thus improving the academic results of their students. Italy's m@t.abel project was also on a national scale, and it was an example of a programme that adopted methodologies and content that adhered to current international educational research. Thus all of the preceding projects were research-based and intended to bring changes in the teaching practices on a national scale in order to impact students' learning.

However, our comparative analyses suggest that when project goals are only decided at the top level, they can be susceptible to political changes, which we will discuss in more depth in the section on constraints.

Focus of the PD programmes

When analyzing the projects in the various countries, we have to take into account their origin and focus, as not all originate as PD programmes for technology integration into mathematics teaching. Some programmes have a focus on developing the mathematical knowledge of their teachers or their pedagogical content knowledge for mathematics teaching (e.g., FUS in Portugal and CIER in Colombia) – that is, teacher training with a focus on mathematics. Such programmes do not necessarily focus on technology integration, although it may be included tangentially. In other cases, the mathematical focus is included more profoundly (e.g., m@t.abel and DiFiMa in Italy). Projects that do centre on the use of digital technologies may do so with a focus on the technical aspects of certain technology (e.g., training in the use of the Internet or a certain app) but with little consideration for the content knowledge and the pedagogical content knowledge. We conclude that this can leave teachers with a lack of know-how about how to implement such technologies for teaching mathematics (and when they do use them, it tends to be only an add-on to the usual teaching that does not allow them to learn to harness the potential of these technologies), as was the case of Enciclomedia in Mexico (Macías-Herrera, 2014). In Portugal, the MINERVA project was implemented near the advent of digital technologies (mid-1980s). Hence, there was an emphasis on technical issues related to the use of technology, although the development of classroom experiences and the development of tasks that took advantage of the potential of technology were already an important part of the project. In contrast, in the case of FUS, the focus was on the implementation of a new syllabus, promoting the use of technology, of different kinds of tasks and of the development of classroom work centred on the students. In PM, the idea was to allow teachers to select the material conditions they needed instead of equipping schools with technologies they might never use so that the teachers were free to choose the focus of the PD most relevant for them.

Our analysis suggests that a "worst-case scenario" is when a technology integration project does not include an adequate accompanying PD programme, and it only focusses on providing schools or students with a particular hardware. This was the case to some extent, in Mexico, of HDT and Enciclomedia (e.g., the provided smartboards were used more as projection screens; Macías-Herrera, 2014). Such cases have shown that this hardware is wasted, and they have led to the recognition that only providing resources is insufficient. In fact, this recognition led to the development of the PM project in Portugal.

Some PD programmes focus more specifically on the integration of technology with mathematics teaching and learning (e.g., EMAT in Mexico) and/or specifically on how to articulate the teacher training with the (math-syllabus) curriculum (e.g., FUS in Portugal). This was also the case of the INTC Colombian project for integrating new technologies to the mathematics curriculum of secondary school (Castiblanco et al., 1999), which included training teachers to harness the potential of digital technologies for their teaching practice (as did the M.Ed. programme in Mexico). In these programmes, although it was important for teachers to learn about the technical aspects of certain tools (hardware/software), there was also emphasis on the following didactical considerations.

Teachers reflected on

- mathematical ideas in the curriculum and how to teach them using digital technology (as a tool for math); and
- how to design contexts and/or tasks using digital technologies for their students in schools.

This privileging of the "teacher's voice" was evident in the cases of PM in Portugal, where the focus was identified based on the teachers' needs and interests, and in Colombia's INTC, where the decision of the specific focus was a negotiation between the teachers enrolled in the programme and the teachers' educators.

The Italian m@t.abel project focussed on new methodologies for teaching and learning mathematics as a response to the weaknesses in students' mathematics skills identified by the international Programme for International Student Assessment of the Organisation for Economic Co-operation and Development survey. By focussing on new methodologies, the use of digital technologies was an important part, both in terms of teaching in the classroom and teacher training. The focus of DiFiMa is also on methodologies, and, in particular, this project takes into account the necessity of teachers to be supported in exploiting technological affordances, creating collaborative contexts for teachers' work where they can share their practice and work collaboratively with and under the guidance of expert teachers and researchers.

Strategies and methods of the PD programmes

Types of PD programmes and the delivery method

Many of the PD projects (e.g., those in Mexico and Portugal) were implemented only in a face-to-face modality. In recent years, however, due to the availability of more accessible and effective online platforms, "blended" distance programmes have also been offered, combining face-to-face and online delivery methods, as in the case of Italy's m@t.abel and Colombia's CIER projects. The format of CIER was three weeks of online activities, one week of face-to-face activities and then two further weeks of virtual work. Also, the availability of MOOCs, in which the interactions amongst trainees and between trainees and trainers occur online and in totally virtual environments, offered the opportunity to enlarge the audience of the courses to a wide national level. This was the case of the DiFiMa programme at the University of Turin in Italy.

In Mexico, some smaller projects that have attempted to blend face-to-face with online participation have not been successful in the distance component, as the Mexican teachers are less familiar with participation in distance courses or CoP and/or lack the time or resources for working online (Miranda & Parada, 2011).

Types of participants and the adoption of the programme

Many of the programmes focussed on in-service teachers. Such is the case for all of the featured Mexican and Portuguese projects and the m@t.abel project in Italy. Colombia's INTC project included both in-service and pre-service teachers (the latter from the universities that led the programmes in each region). In INTC and the Italian DiFiMa, participants could be any interested teachers (so participation was voluntary based on teachers' willingness to be involved in the training process)– the focus of both projects was to develop classroom teaching experiments, but pre-service teachers could be involved and contribute both in the design phase and in the *a posteriori* analysis of the experiences. Also, in Portugal's PM, the enrolment of the teachers was voluntary, with a significant increase on the number of teachers and students involved across the years (GTM, 2019).

Nevertheless, there are projects at a national level in which participation is compulsory (e.g., Mexico's Enciclomedia, Portugal's FUS and Colombia's CPE). In Enciclomedia, teachers in classrooms that were supplied with the Enciclomedia system had to be trained in its use, which was compulsory. Colombian PD programmes have generally been voluntary participation; the CPE project, however, was mandatory for many schools in the country in order to provide digital technologies to schools (e.g., computers, Internet). CPE trained teachers in the basic use of technological tools (e.g., opening and using an email account) and supported the designing, by teachers, of classroom tasks. As a result of this mandatory participation, many schools, teachers and communities in Colombia (many in remote areas) had their first access to digital technologies.

The integration of theory and practice

Some PD programmes are research-based; that is, they take into account the results of research and theories from mathematics education (and other disciplines) to inform their implementation, guide the teacher training and underpin the design of the tasks implemented by the teachers. The Mexican EMAT and the Colombian INTC projects are such examples. In both projects, some materials (e.g., textbooks for students with task ideas) were produced that were derived from research results. In Portugal's FUS, booklets for teachers were written by experts, focussing on mathematical, or more general issues, and discussing theories and the potentials of digital technologies, projects and problem solving. These booklets included some results from the research on the field.

In INTC, research results were used in the design of the PD programme. For example, theories related to problem solving and to the role of mathematical representations in students' learning (Castiblanco et al., 2004). Some years later, the CIER project also included research results and more general theories (e.g., teachers' technological, pedagogical and content knowledge based on the Technological Pedagogical Content Knowledge model from Mishra and Koehler (2006)) were widely discussed with the group of teacher-experts in order to identify what elements of those theories might guide the design of digital resources for teaching. Both Colombian projects also included discussions and reflections amongst the participating teachers to analyze theories, the design of tasks and teaching proposals integrating digital technologies and their implementation. This was also the case in Mexico's smaller M.Ed. programme and in the Italian DiFiMa where participating teachers are introduced to some mathematics education theories to help them understand the constructs that underpinned the design of the teaching activities being proposed to them. This helped them to understand the design choices and supported them in implementing the classroom activities more effectively.

In addition, in the Colombian projects, some of the teachers' implementations in their classes were observed by expert teachers (and other participant teachers in the PD programme) and used for promoting teachers' reflections on their practice and on topics such as the math curriculum. In all the aforementioned projects from the three countries, the theoretical integration also helped the teachers to analyze the results of the teaching activities alongside the teacher educators/researchers and to reflect on their own professionalism. For these reasons, these projects are interesting examples of theory/practice integration within PD programmes.

Structure of the PD program and how teachers are organized

Some of the PD programmes are based on knowledge from experts, assuming that teachers are the recipients of that knowledge. Other programmes consider that teachers can be engaged as active agents in the design and development of the PD programme. In some cases (e.g., PM in Portugal), the teachers identified the area where they feel they need to improve in their professional knowledge and are organized in a CoP for each school. In this model, the teachers from each region could get together each month to seek the support they needed from an expert teacher. Expert teachers are those who receive training with experts on the field of mathematics education and then become experts who can share their knowledge with their peers. In some cases (e.g., Mexico's EMAT, Italy's m@t.abel, Colombia's CIER), these expert teachers were required to train other teachers in a "cascade" model. The teacher training in CIER and EMAT included an interdisciplinary group of experts (e.g., researchers, designers, communicators) who trained a team of expert teachers who would then train other participating teachers. In the case of m@t.abel, the expert teachers had the role of tutors in the online platform. In the Portuguese FUS, teachers were selected as expert teachers to work with the other teachers, guiding the work in the monthly meetings. In some cases (e.g., Portugal's FUS, m@t.abel, EMAT), the expert teachers were mediators between the national syllabus authors and the schools, taking into account the needs of the teachers.

In the case of Mexico, for Enciclomedia, expert teachers were trained at regional centres during a three-day course (with no follow-up training or support) to lead the training of other teachers to learn the contents of the Enciclomedia system. This was a purely technical training with no didactical training. In the case of EMAT, even though the designers requested follow-up and more continuous teacher support, it was denied by authorities, so training was restricted to up-to-a-week workshops, and uptake by teachers was low (only in places where there were self-appointed coordinators and more continuous support was the uptake by teachers more successful).

Many of the Colombian PD programmes were implemented as a result of extensive cooperation between institutions. For example, several government entities (e.g., Ministries of Education, Environment and Telecommunications) and private companies participated in the CPE project. Other programmes such as INTC and CIER, as well as Mexico's EMAT, collaborated with regional universities and had the support and advice of international experts. In the case of EMAT, the most successful implementations, after the PD programmes, were in places where the regional universities offered long-term support (Sacristán & Rojano, 2009). The INTC programme also added support to its participating teachers through national conferences where they could have contact with the PD programme designers, national and international researchers and with other participating teachers. As a product of these conferences, the National Ministry of Education published some books that were distributed nationwide and are still available on their website.

Strategies for teachers' work in the PD

It has been mentioned how some of the PD programmes involved teachers in discussions on related theories and/or classroom implementations. Our analysis suggests that many of the PD programmes (especially those with in-service teachers) involved teachers in the discussion of ideas and classroom experiences. In some cases, this resulted in the development of CoPs; in other cases, it resulted in teachers working collaboratively but with a level of collaboration weaker than the one of a CoP. Portugal's PM was guided by what the teachers wanted to do with their students and included weekly meetings of each school's CoP, to include peer work inside and outside of the school, through which the teachers shared their experiences.

In Colombia's CIER, the strategy was to offer digital resources and materials to teachers so that they could work with them in their mathematics classes and subsequently reflect and share their experiences in CoPs implemented during the PD programme. In Italy's DiFiMa and m@t.abel projects, classroom experiences were also shared and discussed in CoPs (in m@t.abel, in particular, the CoPs combined face-to-face and online training and discussions).

Resources and materials used

For many of the programmes, materials were created either for teachers to use with their students or specifically for teacher training. In all of the Mexican programmes, most of the materials and resources used were designed for use in the classroom by the teachers. The main software used in EMAT includes "universal, open tools" such as spreadsheets, dynamic geometry, logo and graphic calculators that were accompanied by sets of activities (worksheets) for use with students. Teachers were provided with handbooks and books, written by a team of academic experts, containing tasks and worksheets (classified according to the compulsory curriculum) for each tool.

In the case of Enciclomedia, the materials and resources were special digitized versions of the compulsory primary school textbooks with accompanying digital interactive applications (in CD and in a platform). The classrooms were equipped with a computer and a smartboard. Teacher guidebooks described the structure of the interactive system and materials, and teacher training consisted of familiarizing the teachers with the tools and activities in a purely technical way.

In Portugal, some of the programmes related to the development of support material for teachers and some did not. The FUS project had the strongest focus on such support materials, which took the form of booklets for teachers, which have been described previously. The MINERVA project also produced some materials for teachers but additionally developed software specifically for the teaching of mathematics.

In the Colombian cases, most of the PD programmes provided equipment to the participating schools: mathematics computer laboratory rooms (in INTC case), hardware, software, Internet connection and technical support (e.g., in the CPE case) and website-based digital resources for teachers (e.g., in CIER[11] and in CPE). In most cases, the technological equipment was accompanied by textbooks and other complementary resources for teachers (e.g., worksheets, digital books, applications and videos).

Some PD programs have used videos as part of their teacher training strategy (see also van den Bogaart, Drijvers, & Tolboom, 2019). For example, in the CIER project, videos were made and used for two purposes: (1) as part of the resources offered to teachers (through a web page) so that they could use them in their own classes and (2) as resources used by teacher educators to promote teacher reflection and knowledge. In the latter case, some aspects of a lesson study were included. However, this was neither systematic nor recurring. Also, in the Italian DiFiMa, there have been some, not recurring and systematic, cases of the use of videos, conceived as an effective strategy of the teacher training since it allows teachers to reflect on their role through a deep analysis of the outcomes of the teaching activities.

Even though the m@t.abel project ended in 2012, all the materials, developed and used by the researchers and the teachers during the programme, are still available online on the platform managed by the National Agency for School Development and Autonomy on behalf of the Italian Ministry of Education. In contrast, Mexico's EMAT and Enciclomedia materials are no longer online, although many teachers are still using them (over ten years after the programmes ended), and for the new 2018 curriculum, most of the digital technology recommended activities were taken from EMAT.

Constraints, allowances and incentives of the PD programmes

Given the national priorities that underpin many of the projects, the subsequent teacher engagement and outcomes are dependent on the related constraints, allowances and incentives that might support their active involvement. One important factor is the top-level government/political decision regarding the resources allocated to the design and implementation of such programmes. In particular, when projects are decided at the top level, they can be susceptible to political changes and abruptly cancelled before they attain their goals, as happened for EMAT, which was terminated before full expansion, despite its success. Similarly, Enciclomedia was suddenly cancelled after only a few years due to political changes, in spite of its massive expenditure. In Portugal MINERVA, FUS and PM were all politically motivated and required considerable financial support, which made them vulnerable to political decisions – FUS was suspended due to financial constraints, although in this case, most of the goals of the project had been achieved.

Other factors that affect PD programmes include teachers' opportunities to access technologies, as well as teachers' time constraints, which are well-known to impact the integration by teachers of what they have learned in a PD programme (e.g., in terms of the use of technologies) into classroom practice. However, the duration of the PD programme is also relevant. Longer and sustained PD programmes give teachers the opportunity to reflect on the integration of technology in their practice – for instance, through using technology in their classrooms during the PD programme – with possible support from the participating experts and peers in that programme. This is, for example, the case with DiFiMa: the project started in 2008 and has been able to continuously offer a range of evolving opportunities, such as online commented materials, MOOCs, blended courses on specific contents or methodologies for teachers at the lower or high secondary school level, online suggestions and support and periodical conferences in which teachers can also present and discuss their own experiences.

Most PD programmes assume that teachers have access to technology. Hence most of the projects that we have described have equipped schools with the relevant technology (Colombia's INTC, CPE, CIER; Mexico's EMAT and Enciclomedia; and Portugal's MINERVA and PM projects). In some cases (e.g., INTC, EMAT and MINERVA), mathematics laboratory rooms (equipped with hardware and software) were created in the participating schools with accompanying training for teachers. In the case of PM, the schools received the equipment asked for by the teachers after they had conceptualized their specific project.

The type of teacher enrolment is also an important factor. Voluntary adoption of the programme will result in a group of participants with a deeper recognition of the relevance of the programme. In Portugal's FUS, all upper secondary teachers were enrolled, as were all upper primary school teachers in Mexico's Enciclomedia (where some teachers asked to be transferred because they did not want to participate). In contrast, in Mexico's EMAT, the enrolled schools expanded progressively, and teachers' participation was voluntary. This process of progressive inclusion happened also in the case of Portugal's PM where the schools decided if they wanted to participate and how.

Another participation modality that can have an impact on the results of a PD programme is whether it is face-to-face or at a distance. Once again, there are different allowances and constraints related to these two options. Face-to-face can result in a deeper involvement of the teachers and in a stronger feeling of belonging to a group. However, this modality has constraints related to the need to be together in the same place that the distance option does not.

The exclusivity of the time that teachers dedicate to the PD program is also an important factor. In the case of the Colombian CIER project, the participating teachers had a time of exclusive dedication to the program (while the students were on holiday). This enabled the teachers to focus on the program and commit to the training activities. In addition, the program certified their participation and evidenced this on social networks (e.g., Facebook). At an institutional level, the schools recognized teachers' participation as part of the annual evaluation of teacher performance.

Evaluating the impact of the PD programmes

A PD programme can be assessed in a systematic and/or systemic way in order to have information about both its impact on teachers' practice and the effectiveness of its components (e.g., focus, methodology, tools). Evaluating a programme's outcomes is particularly important to learn from previous experiences and improve the design and implementation of future programmes. However, this aspect is often undervalued and disregarded. Many of the PD programmes we considered were not formally evaluated, and, even when this was carried out, the results were not considered as a starting point to design and develop further programmes. For this reason, the impacts of PD programmes are often neither shared nor discussed on the national or international level.

That is the case of Portugal, where the final reports of the projects are often assumed as classified information and not publicly available. Nevertheless, some characteristics of the PD programme are broadly discussed, allowing for the identification of central points to promote or avoid in future initiatives. This is the origin of ideas, such as in FUS, where some teachers were invited to assume an important role in the PD programme as expert teachers. Then in PM, stronger recognition of the role of the teachers evolved, allowing them to organize and decide the focus of the project in their school.

In the case of the Italian m@t.abel, the impact of the programme has been assessed, even if, similar to Portugal, the results are not publicly available. Notwithstanding, we can assume that the basic assumptions, on which the programme had been designed and developed, have been deemed as successful. Indeed, they constitute the basis for further programmes such as DiFiMa. Moreover, the Italian PD programmes have a very strong connection with research activities, and their impacts are also evidenced through the scientific publications that draw data from the teachers' experiences in their classes (e.g., see Taranto, Arzarello, & Robutti, 2017; Loisy et al., 2019).

However, some PD programmes have had an associated evaluation project to determine their impact. Such was the case of Mexico's EMAT programme, where schools, teachers and students were surveyed and observed in order to determine the impact of the project (see Trigueros & Sacristán, 2008). The evaluation noted discrepancies between the use that schools and teachers claimed and what students perceived, which suggested a poor implementation of the programme, perhaps due to insufficient teacher training and support. In addition to an overall evaluation, for EMAT, as well as Enciclomedia and the M.Ed. programme, there were associated academic research projects that analyzed parts of the projects (e.g., see, respectively, Trigueros & Sacristán, 2008; Trigueros & Lozano, 2012; Sacristán, Sandoval, & Gil, 2011; also Trouche et al., 2012).

In the case of the Colombian CPE project, an external evaluation was carried out by another state entity (the National Consulting Centre) to assess its impact, results and sustainability. This adopted a mixed-methods approach (using quantitative and qualitative methods) and took into account the data of a sample of selected schools that participated in the project. A summary of this assessment is available online.[12] In this assessment, it was concluded that the CPE programme managed to positively impact its participating schools (e.g., helping to reduce student dropout and repetition). Another conclusion was that the teachers recognize the relevance of the teacher training, and they want to continue being involved in further actions of technology integration in the classroom. This resulted in the continuity of the project. However, the assessment results suggest that it is necessary to continue investigating which pedagogical methods are most effective to achieve meaningful experiences in the classroom.

Concluding remarks: Towards pragmatic theories for PD concerning technology use in school mathematics

The analysis of the various national experiences involves a "learning over time" that enables us to gain insights that inform and comprise a PD programme for mathematics teachers so that they can exploit technological affordances, be supported and share their practice and work. In this way, we can define better the pragmatic theories underlying teacher PD concerning technology use in mathematics. In our process of analysis, we identified some common aspects that in our conclusions both highlight the interconnections between those different aspects and shed light on the central role of mathematics.

Our pragmatic theories for PD design concerning technology use in mathematics

Focussing on mathematics with technology rather than on technology (for math education)

The experiences revealed that, although most of the teachers expressed the need to be informed and trained about the functionalities of the new technologies, it is by focussing on both the mathematical content knowledge and the pedagogical content knowledge for mathematics teaching that they are able to exploit the opportunity offered by the PD programme and become aware of the limits and potentials of the integration of technologies in their teaching practices (Faggiano, 2009; Clark-Wilson et al., 2014). In addition, our experiences revealed that teachers need to experience mathematical technologies for themselves "as learners". This experience develops teachers' professional instrumental geneses and best practices with digital technology (Haspekian, 2012) and helps them in turn to support their students' instrumental geneses (Trouche, 2004).

Providing schools with digital technology is only part of the solution

Some political decisions in several countries provided materials or hardware (e.g., big initiatives at the end of the 20th century), and these resulted in less than successful interventions, as the resources were left mostly unused (e.g., see Trouche et al., 2012). Our experience reveals that the PD programmes, which also offered teachers training and in situ support (technical and pedagogical), are considered more successful than those simply providing schools with digital technology.

Taking into account teachers' needs

The recognition of the knowledge, the relevance and the needs of the teachers is also an important issue in PD. The enrolment of the teachers is deeper and stronger when they see themselves as a relevant part of the project, having the opportunity to discuss what they think is important and being able to define the project based on their needs, their interests and their classroom reality.

Although the underlying learning theories for particular pedagogical approaches may or may not be made explicit for participating teachers, our conclusion is that the success of a PD program is how it responds to the needs of the specific teachers it is directed to.

Enabling connections between different actors: The teachers, the expert teachers, the researchers and the institutional agency

At the end of the 20th century, partnerships between academia and schools arose but depended on good coordination amongst academic educators, authorities and schools. We have recognized the contributions of experts in mathematics education and their collaboration with schoolteachers (in some cases also with expert schoolteachers) working as authors of the educational activities proposed. This allowed the most interesting and innovative suggestions coming from the international mathematics education research to be collected and dropped into the reality of the school system. In this process, the role of other institutional agencies, such as those responsible for school development and national assessment or the professional councils of mathematicians or teachers, were also found to be important.

Linking teacher training with classroom practice

Professional development opportunities need to include opportunities for the teachers to plan activities for their students during the PD programme. Indeed, through the design, development and analysis of teaching practices with technology made during the PD programme in collaboration with expert teachers and researchers, teachers can develop meta-levels of reflections on both the educational processes and the results of these processes. This would also enable explicit connections to be made with the school curriculum within CoPs and with experts' support.

Ensuring the duration of the programme and support given to the teacher

It takes time for teachers to effect the transformation of mathematics teaching practices with technology both in terms of the timing and frequency of support (e.g., continuous and sustained support). This support includes (but is not restricted to) technical and logistical aspects, didactical aspects (on the use of digital technologies in the classroom) and opportunities for teachers to participate in teacher communities (e.g., through social networks).

Developing and maintaining fruitful CoPs

Shaping CoPs is a useful strategy for teacher training because it allows greater communication between teachers and the sharing of their experiences. However, the CoPs are not easy to implement, and their continuity after the PD program is difficult.

Maintaining political support for the PD programmes

Changes in governments often imply huge changes in educational policies and, in particular, in teacher training programmes (e.g., see Trouche et al., 2012). However, in some cases, it can happen that educational policies give continuity to the PD programme, allowing PDs to be strengthened.

Exploiting the potentials of the newest technological innovations

New technologies are being developed every day, and PD programmes should be flexible to incorporate opportunities to reflect on the new affordances available for learning, teaching and teacher training. This can mean, for example, exploring the potentials of touch and multiple screens, augmented reality and eye tracking. Moreover, there are some other kinds of technological innovations that, in our view, deserve more attention, such as the MOOCs or videos. The latter, for example, can be used not only for classroom experiences to reflect upon but also for teacher training (e.g., Van den Bogaart, Drijvers, & Tolboom, 2019).

Final remarks and call for further investigation

We believe that the process of analysis that we carried out met our goal of beginning to define a useful framework for the design of future PD programmes, as well as for researchers to analyze existing programmes. But it is also a framework that needs to be developed further, validated with further case examples and evolved.

On the one hand, as technology and global communications advance, the strategies of the PD programs may need to adapt (e.g., through online programs or MOOCs). A training strategy that was successful a decade ago may no longer be as useful. So the framework, if it is to be useful, should be adaptable to new contexts and settings.

On the other hand, and more importantly, the development of the framework needs to take into account other experiences from around the world because PD practices tend to align to the context and needs of a particular country or region. We believe that similar investigation from countries with different cultural traditions, such as Anglo-Saxon countries, northern European countries or Asian countries, need to be developed. This would allow our set of pragmatic theories to be deepened and further improved.

Acknowledgement

We are very grateful to Alison Clark-Wilson of the UCL Knowledge Lab, UCL Institute of Education, University College London, United Kingdom, for her ideas and comments on this chapter.

Notes

1 INTC – *Proyecto "Incorporación de Nuevas Tecnologías al Currículo de Matemáticas de la Educación Básica y Media de Colombia"*: https://www.mineducacion.gov.co/1621/article-81040.html
2 CIER – *Centros de Innovación Educativa Regional*: http://aprende.colombiaaprende.edu.co/es/edusitiocier
3 CPE – *Computadores Para Educar*: https://www.computadoresparaeducar.gov.co/
4 m@t.abel - *Progetto m@t.abel*: http://www.scuolavalore.indire.it/superguida/matabel/
5 DiFiMa – *Didattica della Fisica e della Matematica*: https://difima.i-learn.unito.it
6 PLS – *Piano Lauree Scientifiche:* https://www.pianolaureescientifiche.it
7 EMAT – *Enseñanza de las Matemáticas con Tecnología*: https://web.archive.org/web/20140605094110/http://www.efit-emat.dgme.sep.gob.mx/
8 *Enciclomedia:* https://web.archive.org/web/20100725200920/http://www.enciclomedia.edu.mx/ https://www.sep.gob.mx/work/models/sep1/Resource/2959/4/images/LB Enciclomedia.pdf
9 HDT – *Habilidades Digitales para Todos*: https://sep.gob.mx/work/models/sep1/Resource/2959/5/images/LB%20HDT.pdfhttps://sep.gob.mx/work/models/sep1/Resource/2959/5/images/LB%20HDT.pdf
10 MINERVA – *Meios INformáticos no Ensino: Racionalização, Valorização, Atualização*
11 *Capsulas Educativa Digitales:* http://aprende.colombiaaprende.edu.co/sites/default/files/naspublic/ContenidosAprender/index.html
12 Impact of CPE: https://colombiatic.mintic.gov.co/679/w3-article-74005.html

References

Aldon G., Arzarello F., Panero M., Robutti O., Taranto E., & Trgalová J. (2019). MOOCs for mathematics teacher education to Foster Professional Development: Design principles and assessment. In Aldon G., & Trgalová J. (Eds.), *Technology in Mathematics Teaching: Mathematics Education in the Digital Era* (pp. 223–246). Cham: Springer. 10.1007/978-3-030-19741-4_10.

Ball, D., Thames, M., & Phelps, G. (2008). Content knowledge for teaching: What makes it special? *Journal of Teacher Education*, 59, 389–407. 10.1177/0022487108324554.

Castiblanco, A. C., Camargo, L., Villarraga, M. E., & Obando, G. (1999). *Nuevas tecnologías y currículo de matemáticas. Apoyo a los Lineamientos Curriculares*. Bogotá: Ministerio de Educación Nacional.

Castiblanco, A. C., Urquina, H., Camargo, L., & Moreno, L. E. (2004). *Tecnología Informática: Innovación en el Currículo de Matemáticas de la Educación Básica Secundaria y Media*. Bogotá: Ministerio de Educación Nacional.

Clark-Wilson, A., Aldon, G., Cusi, A., Goos, M., Haspekian, M., Robutti, O., & Thomas, M. (2014). The challenges of teaching mathematics with digital technologies-the evolving role of the teacher. In Liljedahl, P., Nicol, C., Oesterle, S., & Allan, D. (Eds.), *Proceedings of the Joint Meeting 1–87 of PME 38 and PME-NA 36* (Vol. 1, pp. 87–116). Vancouver, Canada: PME.

Clark-Wilson, A., & Hoyles, C. (2018). A research-informed web-based professional development toolkit to support technology-enhanced mathematics teaching at scale. *Educational Studies in Mathematics*. 10.1007/s10649-018-9836-1.

Faggiano, E. (2009). Leading teachers to perceive and use technologies as resources for the construction of mathematical meanings. In Durand-Guerrier V., Soury-Lavergne S., & Arzarello F. (eds.), *Proceedings of CERME 6* (pp. 1310–1319). Lyon: INRP – Institut National de Recherche Pédagogique.

GTM – Grupo de Trabalho de Matemática (2019). *Recomendações para a melhoria das aprendizagens dos alunos em Matemática (1.ª versão)* (pp. 182–189). Lisboa: DGE. https://dge.mec.pt/sites/default/files/Curriculo/recomendacoes_para_a_melhoria_das_aprendizagens_dos_alunos_em_matematica.pdf.

Haspekian, M. (2012). *Understanding teachers practices and professional geneses with technology. International Congress on Mathematical Education, ICME 2012*, Seoul, South Korea.

Loisy, C., Sabra, H., Courtney, S. A., Rocha, K., Gracin, D. G., Aldon, G., ... & Robutti, O. (2019). Analyzing teachers' work with resources: Methodological issues. In L. Trouche, G. Gueudet & B. Pepin (Eds.) *The 'Resource' Approach to Mathematics Education* (pp. 257–321). Springer, Cham.

Macías-Herrera, G. (2014). Los docentes ante las políticas enfocadas en TIC (Master Thesis (Maestría en Desarrollo Educativo - Formación: Política Educativa), Universidad Pedagógica Nacional, Mexico City, Mexico. http://xplora.ajusco.upn.mx:8080/jspui/handle/123456789/446.

Miranda, L., & Parada, S. E. (2011). *Están preparados los profesores de matemáticas para implementar las tecnologías digitales en sus prácticas* Presented at the *XIII CIAEM-IACME*, Recife, Brasil. http://ciaem-redumate.org/ocs/index.php/xiii_ciaem/xiii_ciaem/paper/download/1593/317.

Mishra, P., & Koehler, M. J. (2006). Technological pedagogical content knowledge: A new framework for teacher knowledge. *Teachers College Record*, 108(6), 1017–1054. 10.1111/j.1467-9620.2006.00684.x.

Ponte, J. P. (1994). *O projeto MINERVA: Introduzindo as NTI na Educação em Portugal - Relatório do projeto MINERVA*. Lisboa: DEPGEF. Retrieved from http://www.educ.fc.ul.pt/docentes/jponte/docs-pt/94-Ponte(MINERVA-PT).doc.

Sacristán, A. I., & Rojano, T. (2009). The Mexican National Programs on Teaching Mathematics and Science with Technology: The legacy of a decade of experiences of transformation of school practices and interactions. In Tatnall A. & Jones A. (Eds.), *Education and Technology for a Better World* (pp. 207–215). 10.1007/978-3-642-03115-1_22.

Sacristán, A. I., Sandoval, I. T., & Gil, N. (2011). Teachers engage in peer tutoring and course design inspired by a professional training model for incorporating technologies for mathematics teaching in Mexican schools. In Joubert M., Clark-Wilson A., & McCabe M.

(Eds.), *ICTMT 10: Enhancing Mathematics Education Through* (pp. 248–253). http://www.mccabeme.myweb.port.ac.uk/ictmt10proceedings2.pdf.

Taranto, E., Arzarello, F., & Robutti, O. (2017). Analyzing MOOCs in terms of their potential for teacher collaboration: The Italian experience. In *Proceedings of Tenth Congress of the European Society for Research in Mathematics Education – CERME 10* (1, pp. 2478–2485). DCU Institute of Education and ERME.

Teixeira, P. (2004). *O Acompanhamento Local como modelo de desenvolvimento curricular em Matemática*. Coleção Teses: Associação de Professores de Matemática.

Tondeur, J., van Braak, J., Sang, G., Voogt, J., Fisser, P., & Ottenbreit-Leftwich, A. (2012). Preparing pre-service teachers to integrate technology in education: A synthesis of qualitative evidence. *Computers & Education*, 59, 134–144.

Thomas, M. O. J., & Hong, Y. Y. (2013). Teacher integration of technology into mathematics learning. *International Journal for Technology in Mathematics Education*, 20(2), 69–84.

Trigueros, M., & Sacristán, A. I. (2008). Teachers' practice and students' learning in the Mexican programme for Teaching Mathematics with Technology. *International Journal of Continuing Engineering Education and Life-Long Learning*, 18(5/6), 678. doi:10.1504/IJCEELL.2008.022174.

Trigueros, M., & Lozano, M.-D. (2012). Teachers teaching mathematics with Enciclomedia: A study of documentational genesis. In G. Gueudet, B. Pepin, & L. Trouche (Eds.), *From Text to 'Lived' Resources: Mathematics Curriculum Materials and Teacher Development* (pp. 247–263). doi:10.1007/978-94-007-1966-8_13.

Trouche, L. (2004). Managing the complexity of human/machine interactions in computerized learning environments: Guiding students' command process through instrumental orchestrations. *International Journal of Computers for Mathematical Learning*, 9(3), 281–307.

Trouche, L., Drijvers, P., Gueudet, G., & Sacristán, A. I. (2012). Technology-driven developments and policy implications for mathematics education. In Clements M.A., Bishop A. J., Keitel C., Kilpatrick J., & Leung F. K. S. (Eds.), *Third International Handbook of Mathematics Education* (pp. 753–789). 10.1007/978-1-4614-4684-2_24.

van den Bogaart T., Drijvers P., & Tolboom J. (2019) Co-design and use of open online materials for mathematics and science didactics courses in teacher education: Product and process. In Aldon G. & Trgalová J. (Eds.), *Technology in Mathematics Teaching: Mathematics Education in the Digital Era* (13, pp. 205–222). Cham: Springer. 10.1007/978-3-030-19741-4_9.

Appendix

TABLE 4.1 Aspects and sub-aspects of the analyzed PD programmes

		Countries >	*Colombia*			*Italy*		*Mexico*			*Portugal*		
		Projects >	*INTC*	*CIER*	*CPE*	*m@t.abel*	*DiFiMa*	*EMAT*	*Encm*	*M.Ed.*	*MINERVA*	*FUS*	*PM*
Themes	*Aspects*	*Sub-aspects*											
Vision/ goals and policies	What are the political and programme/ initiative goals?	Integrate technology: in general (G) or specifically for mathematics education (ME)	ME	ME	G	ME	ME	ME	G	ME	G	ME	ME
		Promoting change in teachers' practice	X	X	X	X	X	X	X	X	X	X	X
		Deepening teachers' professional knowledge (e.g., MPTK)	X	X	X	X	X	X		X	X	X	X
		Improve student academic results	X						X				X
	Relationship with research	Research findings taken into account for design of PD	X	X		X	X	X		X		X	X
		Integration of PD programme with research project	X	X			X	X		X		X	
	Scale	National (N), regional (R), local (university, school, etc.) (L)	N, L	N, L	N	N, L	N, L	N	N	L	N; L	N	N, L

(*Continued*)

TABLE 4.1 Continued

		Countries >	*Colombia*			*Italy*		*Mexico*			*Portugal*		
		Projects >	*INTC*	*CIER*	*CPE*	*m@t.abel*	*DiFiMa*	*EMAT*	*Encm*	*M.Ed.*	*MINERVA*	*FUS*	*PM*
Themes	*Aspects*	*Sub-aspects*											
Focus	Focus on mathematics	Teacher training with a focus on math (not necessarily technology)				X	X					X	X
		Teacher training articulated to the mathematics syllabus (focus on Mathematical Content Knowledge (MCK) and/or Pedagogical Content Knowledge (PCK) for math teaching)		PCK		X			MCKPCK			X	X
		Training in technical aspects of certain hardware/software. (e.g., focus on technical knowledge; no mathematics).		X	X								
	Focus on technology: teacher training with focus on technology	Training for technological integration (tool for mathematics) – i.e., training to use digital technology in mathematics teaching. Focus on- how some digital tools could be useful for mathematics (T) - developing MPTK	T	MPTK		MPTK	MPTK	T (± MPTK)	T	T, MPTK	T, MPTK	T, MPTK	T, MPTK
	Decision of focus	Focus decided by the teachers enrolled (T) or by outsiders (O)	T, O	O	O	O	T, O	O	O	O	O	O	T

		Countries >	*Colombia*			*Italy*		*Mexico*			*Portugal*		
		Projects >	*INTC*	*CIER*	*CPE*	*m@t.abel*	*DiFiMa*	*EMAT*	*Encm*	*M.Ed.*	*MINERVA*	*FUS*	*PM*
Themes	*Aspects*	*Sub-aspects*											
Strategies and methods	Type of PD programme	Delivery/implementation modality: Face-to-face (FF); Online/distance (D), Mixed (FF and online) (M). If MOOC add D-MOOC or M-MOOC	FF	M	FF	M	M	FF	FF	FF	FF	FF	FF
		Type of participant teachers: in-service (IS), pre-service (PS), mixed (M)	M	IS	IS	IS	M	IS	IS	IS	IS	IS	IS
		Adoption of the programme: progressive (P); volunteer (V); compulsory (C)	P,V	P,V	P, C	P,V	P, C	P, C			V	C	V
	Theory/practice integration	Use of research results and theory for design of PD programme	X	X		X	X	X		X		X	X
		Teachers reflect on the theory/practice integration	X	X		X	X			X		X	X
		The PD programme requires classroom implementation	X	X	X	X		X	X	X		X	X
		PD trainer observes classroom implementations	X	X			X	X					

(*Continued*)

TABLE 4.1 Continued

		Countries >	*Colombia*			*Italy*		*Mexico*			*Portugal*		
		Projects >	*INTC*	*CIER*	*CPE*	*m@t.abel*	*DiFiMa*	*EMAT*	*Encm*	*M.Ed.*	*MINERVA*	*FUS*	*PM*
Themes	*Aspects*	*Sub-aspects*											
	Teacher training strategies	Teachers sharing and reflecting on each other's classroom practices and experiences	X	X		X	X			X		X	X
		Teachers training other teachers		X		X						X	X
		Teachers working collaboratively (CW) or CoP, self-support from teachers (SS), experts and teachers (E-T)	E-T	E-T	CW	CoP	CoP, CW, SS, E-T			CW	E-T	CW, SS, E-T	CoP, SS
	Support or advice	Depth /level: continuous (C), local (L), online (O)	L,	L, O		O	C, L, O	L	L	C, L		C	L
		International support or advice	X	X				X					
		Cooperation between institutions	X	X	X	X		X			X		

		Countries >	*Colombia*			*Italy*		*Mexico*			*Portugal*		
		Projects >	*INTC*	*CIER*	*CPE*	*m@t.abel*	*DiFiMa*	*EMAT*	*Encm*	*M.Ed.*	*MINERVA*	*FUS*	*PM*
Themes	*Aspects*	*Sub-aspects*											
	Teacher training methods and resources	Digital resources:- Making use of what is already available (digital tools/ integrating digital tools with non-digital ones) (A) -Using/presenting/training in new digital resources (N) -Type of use (as content-C, for delivery-D, for teacher support-TS)	A, C	A, N,	A, C	A, N, C, TS	A, N, C, TS	A, C	N, C	A, C	A, N C, D	A, N D	A D
		Use of videos -for teacher training produced by trainers (TT) -by teachers for reflection of their own practice (or of students) (S) for discussion with peers (P)		S, P			S						
		Development of some kind of technology (app, digital textbook, videos, etc.) by experts (E) for teachers' use in their classrooms or teachers (T) involved in producing technology (apps)	E	E	E				E		E		
		Production of school materials (activities) integrating technology by experts (E) or by teachers (T)	T	T	T	E	E,T	E	E	T	E,T	E,T	T

(*Continued*)

TABLE 4.1 Continued

		Countries >	*Colombia*			*Italy*		*Mexico*			*Portugal*		
		Projects >	*INTC*	*CIER*	*CPE*	*m@t.abel*	*DiFiMa*	*EMAT*	*Encm*	*M.Ed.*	*MINERVA*	*FUS*	*PM*
Themes	*Aspects*	*Sub-aspects*											
	Assessment	Development of skills and competences of teachers related to the students' assessment		X								X	
		Teachers' self-assessment	X	X	X					X		X	X
Constraints or allowances and incentives	Government and/or institutional policies (context)	Impact of government/ institutional policies on design (D), continuity of programme (C), availability of resources (RA), etc.	C, CA	D, C, RA	C, RA			C, RA	D, C, RA		D, C, RA	D, C, RA	D, C, RA
	Access	Level of access that teachers have to technology (availability of resources): low (L), medium (M), high (H)	M	H	L			M	H	L	M	M	H
	Time	Length of programme (in years: ½, 1, 2, etc)	2	1	2	6	11	8	4	5	10	3	6
		Teachers' time for involvement in programme: low (L), medium (M), high (H)	H	H	M	M	M	L	M	M	M	H	H
		Teachers' time for implementation in class: low (L), medium (M), high (H)	M	H	L	M	M	L	L	M		H	H

		Countries >	Colombia			Italy		Mexico			Portugal		
		Projects >	INTC	CIER	CPE	m@t.abel	DiFiMa	EMAT	Encm	M.Ed.	MINERVA	FUS	PM
Themes	*Aspects*	*Sub-aspects*											
	Affect/ motivation/ perceptions	Did the programme seek to include aspects deemed to be motivating for teachers (T) and/or for their students (S) to participate and learn?		T			T	T, S	S	T	T	S	T, S
		Did the designers/ trainers make explicit to participants their perceptions on the role and use of digital technologies in math classes?	X			X	X	X		X	X	X	
		(If info is available – see assessment): Did teachers' perceptions regarding the use of DT in mathematics classes affect their participation in the programme and/ or implementation of what was learned in their practice? Yes (Y), partially (P), no (N)				Y	P	Y		Y	Y	Y	Y

(Continued)

TABLE 4.1 Continued

		Countries >	*Colombia*			*Italy*		*Mexico*			*Portugal*		
		Projects >	*INTC*	*CIER*	*CPE*	*m@t.abel*	*DiFiMa*	*EMAT*	*Encm*	*M.Ed.*	*MINERVA*	*FUS*	*PM*
Themes	*Aspects*	*Sub-aspects*											
Evaluating the impact (of the PD programme)	Systematic and systemic evaluation of the programme	Was there an evaluation of its impact on practice (evaluation of the teachers within and after the programme)? Yes (Y), partially (P), no (N)	P			P	P	P		P	N	N	N
		Were the components of the training programme evaluated? Yes (Y), partially (P), no (N)	P	P		P	N	P			P		
	Proposals for improvement	Were proposals for improvements for future implementations made? Yes (Y), partially (P), no (N)	P	Y			P	P			P		

5

TECHNOLOGY-RICH ASSESSMENT IN MATHEMATICS

Maria Fahlgren,

KARLSTAD UNIVERSITY, SWEDEN

Mats Brunström,

KARLSTAD UNIVERSITY, SWEDEN

Frederik Dilling,

UNIVERSITY OF SIEGEN, GERMANY

Bjarnheiður Kristinsdóttir,

UNIVERSITY OF ICELAND, ICELAND

Guido Pinkernell and

HEIDELBERG UNIVERSITY OF EDUCATION, GERMANY

Hans-Georg Weigand

UNIVERSITY OF WÜRZBURG, GERMANY

Introduction

Assessment is an important part of the teaching and learning of mathematics. Around the world, innovative projects on the use of technology in assessing mathematics are currently being run. The literature concerning technology-rich assessment is rapidly growing (e.g., Aldon, Cusi, Morselli, Panero, & Sabena, 2017; Drijvers et al., 2016) and has already identified a range of aspects. The chapter aims, first, to create a framework incorporating what we consider to be important aspects of technology-rich assessment in mathematics that have been identified in this literature. These aspects are theoretically explicated in the first part of the chapter. The chapter then uses this framework to analyze and discuss three innovative approaches to technology-rich assessment which have attracted considerable attention and are distinctive in their character: STEP (Seeing The Entire Picture) in Israel, SMART (Specific Mathematics Assessments that Reveal Thinking) in Australia and STACK (System for Teaching and Assessment using a Computer algebra Kernel) in the

United Kingdom. This second part of the chapter aims, then, to provide a systematic analysis (based on the created framework) to capture the main characteristics of these approaches and highlight some differences between them.

Aspects of technology-rich assessment

This section introduces the aspects which constitute the framework and then examines each of them in more detail. Beyond the most obvious classification in terms of *summative* and *formative* assessment, some new aspects of digital assessment have been highlighted in the literature: assessment *with* as opposed to *through* technology depending on whether technology is used as a mathematical tool or as a testing environment, *migratory* as against *transformative* approach according to the role of technology in relation to task features, the different *functionalities of technology* emphasizing different ways that technology can support various agents and the various aspects of *feedback*, which is one of the most effective forms of support for learning, not only in digital environments (see Hattie & Timperley, 2007).

Summative and formative assessment

Usually, assessment in mathematics is associated with the evaluation of student knowledge and learning, most often at the end of a course or teaching unit, with the aim of grading students' level of proficiency. This form of assessment is often referred to as *summative*. In contrast, *formative* assessment aims at gathering information about a student's current state of knowledge and learning to use them as guidance for teachers and students in subsequent teaching and learning (e.g., Stacey & Wiliam, 2013). Digital technology has been used to enhance both summative and formative assessments in various forms.

Assessment with and through technology

One major characteristic of technology-assisted assessment is "assessment *with* technology and assessment *through* technology" (Drijvers et al., 2016, p.12). An example of assessment with technology is a (traditional) paper-based assessment where students have access to "mathematically able" software as tools for problem solving (Stacey & Wiliam, 2013). In assessment through technology, on the other hand, technology serves as a testing environment, e.g. in The Programme for International Assessment (PISA), computer-based versions of the test have been offered since 2012 (Jerrim, 2016).

Stacey and Wiliam argue, "The context of assessment should not differ significantly from the context of instruction" (2013, p. 737). If doing mathematics involves the use of various types of mathematical software, such as CAS (Computer Algebra System), graphing tools or dynamic geometry environments (DGE), these tools must be available to students during assessment as well. To achieve this, a digital assessment system needs to provide students access to these types of software. Drijvers (2018) describes an example of an assessment system, the Finnish Abitti, that offers students the possibility of using mathematical software that they are familiar with. Abitti is an examination system that boots from a USB-stick or a virtual machine, allowing access to different CAS systems. This is an example of assessment done both with and through technology.

Migratory and transformative approach

The use of digital tools in assessment raises the following two questions: How do the examination tasks change? To what extent does the change of media influence what students do (mathematically) during the assessment? (Jerrim, 2016; Threlfall, Pool, Homer, & Swinnerton, 2007). To address this, Ripley (2009) introduces the distinction between *migratory* and *transformative* (or transformational) approaches to support assessment by using technology. In a migratory approach, traditional paper-based tasks are transferred into digital formats. There could be several reasons for doing this – e.g., to deliver a test more easily and to gain an automated generation of test items, feedback and scoring. Perhaps the most obvious example of a migratory approach is when a multiple-choice test changes medium from paper and pencil to a digital format.

If the intention is to use digital technology to create what Ripley denotes as "innovative assessment", then the nature of the tasks should differ from those usually used in traditional tests to obtain "skills and abilities not normally assessed" (Ripley, 2009, p. 94). This approach to assessment, referred to as transformative, might influence curriculum design and lead to educational change.

Laborde (2001) identified four different task categories in a study on teachers' design of tasks for a DGE. One category consists of tasks in which the DGE is "facilitating the mathematical task" (p. 293). In these tasks, the DGE could be used as a visual amplifier (e.g., helping students to identify important properties). However, the DGE is not really used as a problem-solving tool. Laborde (2001) is using the words of Pea (1985), arguing that "in this kind of use for visualisation, technology was simply an *amplifier* and *not a reorganiser*" (p. 289). Thus we regard this kind of task as non-transformative. For a transformative approach, there must be some element of reorganization. The following is an example of a task that we regard as transformative.

> Let P be the extreme point of the function $f(x) = x^2 + bx$, where b is a real number. Make an appropriate construction in a dynamic mathematics software (DMS) environment and study the position of point P for different values of the parameter b. Make a conjecture. Investigate if your conjecture can be generalized.

In this task, the students are requested to use a DMS environment to investigate a mathematical situation by searching for a pattern and to formulate and generalize a conjecture (Fahlgren & Brunström, 2014). We regard this example as transformative since technology is (almost) necessary to solve the task, and it requires reasoning not assessed in traditional tests. According to Ripley, there is a need for more emphasis on the transformative approach. Digital technology could increase the opportunities to assess important mathematical competencies (e.g., problem solving and reasoning) and hence to increase the focus on practicing these abilities.

To sum up, while assessment through (and not with) technology implies a non-transformative approach, assessment with technology could be transformative depending on the type of task used.

Functionalities of technology

The Formative Assessment in Science and Mathematics Education (FaSMEd) project was a three-year international design research project focussing on the use of technology to facilitate formative assessment in mathematics and science classrooms (Wright, Clark, & Tiplady, 2018).

The main contribution to research on the FaSMEd project is a three-dimensional framework that connects (a) five key strategies for formative assessment (in short): clarifying goals, engineering constructive classroom work, providing feedback, students as peers for other students and students as owners of their own learning (see Black & Wiliam, 2009); (b) three agent roles (teacher, student and peers); and (c) three functionalities of technology defined by researchers in the FaSMEd project based on FaSMEd participants' experience (Aldon et al., 2017). The three defined functionalities of technology are (1) sending and displaying, (2) processing and analyzing and (3) providing an interactive environment.

The first functionality refers to how technology can support communication and discussions between agents (e.g., by displaying students' work on a screen or sending messages and files). The second functionality refers to data collection and analysis made possible by technology (e.g., automatically generated feedback or statistical overviews and tracking of students' learning paths). Finally, the third functionality refers to how technology can support both individual and collaborative work on mathematical/scientific tasks (e.g., explorative learning environments, shared dynamic worksheets and spreadsheets; Cusi, Morselli, & Sabena, 2017).

Aspects of feedback

Giving feedback is an essential mechanism through which a learner receives guidance based on an assessment of his/her state of knowledge. Without feedback, a learner would be at loss as to whether and how to continue and what to concentrate on. Hattie and Timperley (2007, p. 81) define feedback as being "conceptualized as information provided by an agent (e.g., teacher, peer, book, parent, self, experience) regarding aspects of one's performance or understanding". Their model of effective feedback is based on the implicit assumption that feedback is adapted to the individual needs and expectations of the learner. Hence, a distinct educational purpose comes in by regarding feedback as "Advice for Action" (Ras, Whitelock, & Kalz,

2015), for which the content of feedback needs to be adjusted to various characteristics of the learner. Narciss (2008) describes three parameters where adjustment can take place: (1) function (cognitive, e.g., knowledge regarding the learning object; metacognitive, e.g., learning strategies; and motivational, e.g., perception of competence), (2) content (components addressing the various aspects of knowledge, e.g., correctness, task constraints, errors made, how to proceed) and (3) form and mode (e.g., adaptivity and modality in feedback presentation, timing and number of tries after feedback).

Which of the three parameters supporting effective adjustment depends on the characteristics of the learner. For learners who already have significant experience with the mathematical content being assessed, the mere information about correctness or obvious mistakes can be sufficient to revise and correct the answer (Johnson & Priest, 2014; Ras et al., 2015). Beginners, on the other hand, need feedback of an explanatory and supportive kind, such as scaffolding or even worked out examples (Rittle-Johnson & Koedinger, 2005). With regard to the learner's overall level of mathematical achievement, Shute (2008) suggests that information for low achievers should be provided promptly, while information should be delayed for high achievers. While the potential of automatic feedback for self-regulated learning with digital learning environments has repeatedly been pointed out (e.g., Mory, 2004), research on the effects of automatic feedback on the learning of mathematics seems just to be starting. In a study on how human feedback compares with automatic feedback, Livne, Livne and Wight (2007) reported that their automatic scoring system was better in identifying error patterns than humans. Olsher, Yerushalmy and Chazan (2016) expect that automatic assessment could help teachers to focus their feedback on aspects of understanding rather than error correction, thus pointing towards an educational effect that automatic assessment can have on teachers. Other studies compared the effects of different types of feedback on knowledge gain, motivation or other variables (e.g., Narciss et al., 2014; Pinkernell, Gulden, & Kalz, 2019).

Categories for analysis

A. **Localization** of feedback: micro-level (i.e., task level) or macro-level (i.e., overall performance) and levels in between
B. **Receiver** of feedback: i.e., to whom the feedback is addressed (e.g., learner, teacher, parents)
C. **Content** of feedback (e.g., right/wrong, hints or error information or worked out examples) and the **form** in which feedback is given (e.g., timing or language)
D. **Adaptivity** (building on the aspects of content or form, as specified earlier) to levels or aspects of knowledge that become apparent in students' answers within formative or summative assessment or that follow personal traits or preferences of students that could result in different feedback content or form.

Analysis of assessment systems

This section uses the framework we have set out in the previous section to describe and analyze three different assessment systems. The analysis is not based on empirical investigations; instead, existing papers have been used to analyze the systems. The main reason we choose to focus on these three systems is that they display different aspects of digital assessment that we consider important for future development. Moreover, the three approaches are all innovative and have been extensively described in the English-language literature.

STACK

STACK is a software platform for authoring computer-aided assessment (Sangwin, 2013). While it is being used in various mathematical fields, it is the fact that STACK is flexible for implementing various didactical ideas of assessing, understanding and giving feedback that makes it interesting for this review. How this is achieved can be seen from the software's authoring interface for programming a single task. First, the text of a task can contain various forms of representations of mathematical content (e.g., numeric, algebraic or geometric), including JAVA-based interactive elements, which all can be randomized. This means that each time a task is newly called, it comes with different numeric, algebraic or geometric values, or even words. Second, the feedback is based on an automatic CAS-based analysis of the learner's response that follows various predefined categories of mathematical properties (e.g., well-known errors or forms of correctness).

Since STACK is an authoring system, the quality of a STACK-based assessment system depends on the didactical concepts that guide each author through formulating tasks, response analysis and feedback. Hence our analysis is based on the potential of STACK as an authoring system accompanied by selected projects.

Analysis of STACK

The following analysis focusses on publications that directly examine the possibilities and limitations of STACK as an authoring platform. Other publications that report on projects in which STACK is used, but where it is not the object of the research, were omitted from the analysis.

Formative and summative. STACK can be used for summative assessment, but the actual potential lies in the possibilities for formative assessment (i.e., eliciting evidence of student understanding and providing feedback that moves learners forward; cf. Black & Wiliam, 2009). With regard to conceptualizations of mathematical content understanding that rely on representational (Duval, 1999) or operational flexibility in a Piagetian sense (Aebli, 1970), STACK allows various types of representation, as well as open-answer types (Kinnear, 2019). For example, a student could be required to give examples of functions or geometrical figures that satisfy given properties.

With and through. Any assessment system based on STACK is an example of assessment through technology. STACK also allows for implementing interactive geometrical tools. Hence whether it belongs to the category with technology depends on the user and what STACK assessment settings he or she decides to use.

Migratory and transformative. It is possible to use STACK for assessment in both migratory and transformative ways. As an example, where STACK is used in a migratory approach, Sangwin and Köcher (2016) directly transferred paper-and-pencil tests into the STACK environment with the purpose of analyzing whether automatic assessment could replace assessment done by humans. A more transformative type could be open-answer text fields or interactive geometry applets, as mentioned earlier.

Functionalities of technology. As to the functionality "sending and displaying", STACK enables the teacher to review the solutions of each student. Concerning "processing and analyzing", automatic feedback can be generated from the solutions of the students. And as to "providing an interactive learning environment", this can be created by integrating various other tools for representing content and enabling communication between the students.

Aspects of feedback. As a plugin for the learning platforms Moodle or Ilias, STACK provides all the facilities of the parent system for a macro-level assessment (i.e., assessing overall performance). However, its main advantage lies in its potential for conceptualizing a differentiated feedback on a micro-level. For each task, one can define the relevant mathematical properties that the student's answer will be checked for. Furthermore, one can define feedback content of various forms that depend on the outcome of the analysis and explicitly refer to numerical, graphical or algebraic information in the students' answers. Moreover, various types of feedback can be implemented to, for example, provide for the different needs of low and high achievers or learners with other characteristics (Pinkernell et al., 2019). The feature of adaptive testing is not a generic part of STACK but can be added with extra software (Bruder et al., 2017). The receiver of the feedback is the learner, but the information can also be transmitted to the teacher.

STEP

The online assessment platform STEP[1] is currently being developed at the Mathematics Education Research and Innovation Center at Haifa University in Israel. STEP is intended to offer an interactive assignment environment that automatically analyzes student responses and provides meaningful information both to the students and to the teacher. In the interactive environment embedded in the platform, students can perform enquiry-based activities while working on open-ended tasks. Of course, not all types of tasks can be utilized for digitized assessment, particularly when it comes to assessing students' mathematical reasoning. Olsher et al. (2016) suggest that asking students to construct "examples" to justify their answers is one way to tackle this. Besides having a prominent role in the justification of mathematical claims, examples can be checked automatically (Haddif & Yerushalmy, 2015).

One main issue addressed by the STEP research team is the challenge for teachers to survey a great number of student responses in real time and immediately use them as a basis for subsequent instruction. The idea is to off-load teachers from the time-consuming work of collecting and categorizing the student submissions. To do this, the system must go beyond just categorizing answers as being right or wrong. Instead, the categorization must reveal the students' mathematical reasoning, both individually and as a whole group. This provides the teacher with an overview of the distribution between predefined categories, as well as a possibility to identify unexpected responses. The aim is to present information "helping teachers understand the nature of the space of examples submitted by many students" (Olsher et al., 2016, p. 15).

Analysis of the STEP platform

Formative and summative. The STEP platform can be used both to support summative and formative assessments. However, the platform is primarily developed to support teachers in their formative assessment practices by helping them to survey all student responses in real time. In this way, the teachers get information about their students' understanding to use when deciding on the next teaching steps. The platform could also provide students with formative feedback that promotes their reasoning process (Harel, Olsher, & Yerushalmy, 2019).

With and through. Assessment using the STEP platform can be performed both through and with technology. It is an example of assessment through technology since the students submit their answers on the platform. However, the platform can also be used as an interactive environment. For instance, an interactive diagram constructed in GeoGebra can be embedded in the platform. Hence, it also fulfils the requirements for assessment with technology.

Migratory and transformative. Studies using the STEP platform provide several examples of transformative types of task. One example is the following task:

> A, B, C, and D form the quadrilateral ABCD. They are all dynamic and can be dragged. If possible, create 3 examples that are as different as possible from each other, in which the perpendicular bisectors to the sides of ABCD meet in a single point. In the dialogue box formulate a conjecture as general as possible of the conditions in which this happens.
>
> (Harel et al., 2019, p. 3)

Students are asked to use an interactive environment to investigate and provide conjectures about a mathematical relationship. This task could hardly be done by using only paper and pencil.

Functionalities of technology. All three functionalities of technology are at play in the STEP platform. "Sending and displaying" is central since the tasks are sent to the students via the platform; the student responses are sent to the teachers via the platform; the platform displays useful information about students' working methods for the teachers; the platform offers teachers the possibility to select and display information (e.g., different response categories) for the class. The "processing and analyzing" functionality is prominent in the STEP platform since it is developed to automatically process and analyze a large set of student responses. Finally, the platform "provides an interactive environment" since an interactive diagram (e.g., a GeoGebra applet) can be embedded in the platform.

Aspects of feedback. Primarily, the STEP platform provides the teacher with feedback on their students' mathematical thinking. However, in later studies, the focus has also been on the students as the receivers of feedback (cf. Harel et al., 2019). Concerning the localization of feedback sent to the teachers (or the students), the focus is on the task level, even if the undertaking to help teachers track student progression across tasks is identified as a long-term challenge (Olsher et al., 2016). Another interesting feature in relation to localization is the endeavour to provide teachers with useful information at the group level (e.g., whole-class level). The content of the feedback that teachers receive in real time consists of the complete set of student responses, as well as subsets relating to different mathematical characteristics. It could, for instance, be a Venn diagram demonstrating the student solution space by displaying the number of responses in three different categories (Olsher et al., 2016). Concerning adaptivity, it is the teachers undertaking to use the information provided by the platform in the best possible way. An automatic adaptivity is not integrated in the platform.

SMART[2]

The "SMART" Test System has been in development at the University of Melbourne since 2008 (Stacey, Price, Steinle, Chick, & Gvozdenko, 2009). SMART is an online assessment system intended to improve the mathematics learning of upper primary and junior secondary school students. It gives teachers an informative diagnosis of students' conceptual understanding. SMART "is programmed to provide teachers with an automated diagnosis of their own students' stages of development in specific topics and to report on an individual's errors and misconceptions, in order to inform teaching" (Stacey, Steinle, Price, & Gvozdenko, 2018, p. 241).

SMART test questions are related to mathematical problems (e.g., factorization or finding prime numbers) or problems in a real-world context. The response formats for SMART test questions include simple responses using sliders, positioning images and multiple-choice answers. Quite often, either students have to select the answers from a drop-down menu or drag-and-drop formats are used. The questions are usually quick to answer and do not require much working out or long texts.

SMART consists of around 130 tests that cover more than 60 topics from grade 5 to 9 (11- to 16-year-old students,[3] e.g., number and operations, algebra, space, measurement, chance and data). Students usually finish a SMART test on one topic in less than ten minutes.

The guidelines of the SMART tests are (Stacey et al., 2009, p. 4)

- short and easy to administer online tests that can diagnose to what extent students understand the concept;
- informative and prompt feedback to teachers about class and individual performance;
- targeted teaching suggestions that address the conceptual hurdle; and
- strong research backing for the diagnosis and advice given.

Each SMART test is individually evaluated through an intelligent environment created by the authors (Stacey et al., 2009). The tests identify students' understandings, misconceptions and gaps in knowledge, and they provide useful information for teachers about possible future steps, next tasks to solve or instructions to improve students' competencies in a special field.

Analysis of the SMART system

Formative and summative. SMART is a formative assessment system. It provides teachers with information intended to influence the teaching of a certain topic, and it is not intended as a complete assessment of the topic. It diagnoses students' thinking while evaluating students' responses and asks whether the answers are "'wrong in a specific way' rather than just 'wrong'" (Stacey et al., 2018, p. 246). The system assigns the answers to (usually) four different developmental stages. These stages are developed on an item level; they explicitly analyze the content of the question.

With and through. SMART is an example of assessment through technology. Students' answers are evaluated through the system. The system classifies especially students' understanding into different levels or stages.

Migratory and transformative. The questions of the SMART system are mainly quite "classical" questions and emphasize the migratory aspect of assessment. The questions used are related to the results of research in the considered field of interest.

Functionalities of technology. SMART is a prototype of using two of the three defined functionalities of technology: students' answers are sent to the system; they are automatically evaluated and classified, and the results are displayed. Even if the response format in some cases includes interactive elements (e.g., using sliders to place a number on a number line), SMART is not a system which provides a real interactive environment.

Aspects of feedback. Only the teacher receives the results of the test because SMART sees the teacher as the initiator of the developmental process. "We want teachers at the centre of the diagnostic process, because we believe that substantial teacher input is required to overcome most of the conceptual obstacles identified" (Stacey et al., 2018, p. 253). The teacher gets feedback on an intermediate level between micro and macro in the sense that it is based on a number of items, although in a special (small) domain. The teacher gets feedback concerning the developmental stages of the student. Usually, there are four stages, and they describe content-related competencies of a student's results. Moreover, misconceptions and common errors are listed. The interpretation of the feedback seems to be the biggest problem while using SMART. Guzmán Guzmán (2014) did an online survey with 138 teachers on their views about SMART. What she defines as the main problem with SMART is that "there is evidence that suggests that some teachers have difficulties in understanding how the students' diagnoses can be used for teaching" (p. 30). SMART gives suggestions to teachers on how to react to the results of the tests. However, it is still an open problem as to how teachers can react to individually given results of the SMART tests. And, finally, teachers have to evaluate their reactions to students' learning. SMART is a non-adaptive system. The authors wanted to make the system easy and efficient to use for teachers.

Conclusion

This chapter has highlighted several aspects of technology-rich assessment that might be useful to take into account in contrasting and comparing different assessment systems. Nowadays, it is well established that formative assessment has great potential to promote students' learning, and our three examples all illustrate that technology can support this. To understand their various ways of tackling this issue, we compare them in terms of the particular aspects included in the framework that we have used in this chapter.

In SMART, the feedback, which is sent only to the teacher, is based on the overall pattern of each student's responses to a selection of (*non-transformative*) tasks (i.e., at a level *intermediate* between *micro* and *macro*). It is the research-based selection of tasks and task performance analysis that makes it possible for the system to provide information to the teacher on each student's mathematical thinking. In STEP, on the other hand, the feedback is at the task (*micro-*) level. Accordingly, the tasks must be designed to make it possible to categorize students' responses (on the task level) so as to inform the teacher about students' thinking. Most often, the key for STEP to achieve this is assessment *with* technology using a *transformative* approach. In their differing ways, both SMART and STEP provide teachers with feedback on students' performances that goes beyond simply ascertaining the number of successfully solved tasks by revealing students' mathematical thinking. However, it is left to the teachers to use this information to create a formative classroom practice. In comparison to SMART and STEP, the *receiver* of the feedback provided by the STACK platform has predominantly been students. Although STACK has mostly been used for summative assessment, there are features that allow a formative (self-assessment) practice. For example, STACK can provide immediate feedback at the task (*micro-*) level that helps students to improve their performance. These tasks have predominantly been *non-transformative*; however, there are examples in later work of more *transformative* types of tasks.

The results of the analysis of the three assessment systems show that formative assessment is only the beginning of a process. "Making formative assessment easier through online tools should promote its use, but it also seems important to have professional development showing its advantages and distinctive features, and to provide teachers with advice on implementation" (Stacey et al., 2018, p. 259).

Concerning the functionalities of technology, "sending and displaying" and "processing and analyzing" are prominent in all three systems, although in different ways. The third functionality, "providing an interactive environment" is essential in STEP due to the (transformative) type of tasks used. While it is possible to integrate an interactive software into the STACK platform, the SMART system does not provide a genuine interactive environment.

Reflection on the framework

The framework was useful to describe and compare the three assessment systems. However, when doing the analysis, we found some limitations of the framework. Consequently, it could be improved to better capture characteristics of assessment systems, as well as differences between different systems. In the following, we will discuss some of these limitations.

Concerning the migratory and transformative approach, we found it necessary to use "non-transformative" several times since most often an approach is neither wholly migratory nor transformative (e.g., tasks where the technology serves as a visual amplifier and not a reorganizer). One alternative might have been to use the Replace Amplify Transform (RAT) framework for assessing technology integration. In this framework (not specifically developed for assessment), Hughes, Thomas and Scharber (2006) distinguish between technology used to replace, amplify or transform practice.

When it comes to the three functionalities, they are useful at a certain level to describe the assessment systems. However, to pinpoint the differences between systems, the functionalities need to be more fine-graded. For example, an environment could provide interactivity to varying degrees. When describing the SMART platform, Stacey et al. (2018) argue, "Sliders provide a very flexible interactive format" (p. 245), referring to an example where students are asked to place a number on a number line using a slider. In this task, the slider is only used by the students to deliver an answer; the system is not specifically designed to provide students with useful feedback. In our analysis, we regarded this as not being a "real interactive environment". In the STEP platform, on the other hand, dynamic mathematical applets could be embedded in the tasks, offering a high level of interactivity. Accordingly, a more fine-graded framework that captures this difference could have been helpful. However, this difference between SMART and STEP is partly captured in our analysis by the migratory/transformative aspect.

A further example where a more fine-graded framework could have been helpful concerns the feedback aspect "localization". We found "micro" useful for distinguishing between whether the feedback relates to a particular task. In contrast, "macro" offers different scales of operationalization, and it is not at all obvious when a system should be regarded as giving feedback on the macro-level. When analyzing SMART, we found just using micro and macro as inappropriate, and, therefore, we used the idea of an "intermediate" level.

To summarize, a more fine-graded framework could be useful to pinpoint some characteristics and differences between technology-rich assessment systems. However, to be useful for practice, it is important that the framework is not too complex and fine-graded.

Further thoughts

Assessment and feedback are based on tasks and the analysis of task responses. The automatic processes of performance analysis are based on algorithms that have been devised by humans. Also, the selection of tasks for the pool and the formulation of feedback text are done by humans, who must follow some didactical idea or impulse when deciding which tasks are good for assessment and what aspects feedback should focus on. Uncovering these ideas was not included in our analysis, even though they form an essential part of characterizing assessment or feedback. Regarding concepts of digital assessment and feedback that are based on comparing student and expert models during the runtime (cf. Brusilovsky, 2001), we suggest that future discussion of digital systems for mathematics learning should put more emphasis on a theoretical reflection of the didactical concepts that went into the planning of a digital assessment system.

Notes

1 www.visustep.com
2 Specific Mathematics Assessment that Reveal Thinking: http://www.smartvic.com
3 Specific Mathematics Assessment that Reveal Thinking, *Smart tests by topic and year level:* http://www.smartvic.com/smart/usesmart/range.htm

References

Aebli, H. (1970). Piaget and beyond. *Interchange*, 1(1), 12–24.

Aldon, G., Cusi, A., Morselli, F., Panero, M., & Sabena, C. (2017). Formative assessment and technology: Reflections developed through the collaboration between teachers and researchers. In G. Aldon, L. Bazzini, & U. Gellert (Eds.), *Mathematics and Technology* (pp. 551–578). Cham, Switzerland: Springer.

Black, P., & Wiliam, D. (2009). Developing the theory of formative assessment. *Educational Assessment, Evaluation and Accountability*, 21(1), 5–31. doi:https://doi.org/10.1007/s11092-008-9068-5.

Bruder, R., Feldt-Ceasar, N., Kallweit, M., Krusekamp, S., Neugebauer, C., Schaub, M., & Winter, K. (2017). *Digital diagnostic testing tasks (DDTA) – theoretical design and interactive example.* In B. Kaur, W. K. Ho, T. L. Toh, & B. H. Choy (Eds.), *Proceedings of the 41st Conference of the International Group for the Psychology of Mathematics Education*, Singapore (Vol. 2).

Brusilovsky, P. (2001). Adaptive hypermedia. *User Modeling and User-Adapted Interaction*, 11(1–2), 87–110. doi:https://doi.org/10.1023/a:1011143116306.

Cusi, A., Morselli, F., & Sabena, C. (2017). Promoting formative assessment in a connected classroom environment: Design and implementation of digital resources. *ZDM*, 49(5), 755–767. doi:https://doi.org/10.1007/s11858-017-0878-0.

Drijvers, P. (2018). Digital assessment of mathematics: Opportunities, issues and criteria. *Mesure et évaluation en éducation*, 41(1), 41–66. doi:https://doi.org/10.7202/1055896ar.

Drijvers, P., Ball, L., Barzel, B., Heid, M. K., Cao, Y., & Maschietto, M. (2016). *Uses of Technology in Lower Secondary Mathematics Education: A Concise Topical Survey*. New York, NY: Springer.

Duval, R. (1999). *Representation, vision and visualization: Cognitive functions in mathematical thinking: Basic issues for learning*. In F. Hitt & M. Santos (Eds.), *Proceedings of the Twenty-first Annual Meeting of the North American PME Conference*, Cuernavaca, Morelos, Mexico (pp. 3–26).

Fahlgren, M., & Brunström, M. (2014). A model for task design with focus on exploration, explanation, and generalization in a dynamic geometry environment. *Technology, Knowledge and Learning*, 19(3), 287–315.

Guzmán Guzmán, M. A. (2014). The SMART test system: Teachers' views about this formative assessment for mathematics. Master Thesis, Melbourne, Australia: University of Melbourne.

Haddif, G. N., & Yerushalmy, M. (2015). *Digital interactive assessment in mathematics: The case of construction e-tasks*. In K. Krainer & N. Vondrová (Eds.), *Proceedings of the Ninth Congress of European Research in Mathematics Education*, Prague, Czech Republic, (pp. 2501–2508).

Harel, R., Olsher, S., & Yerushalmy, M. (2019). *Designing online formative assessment that promotes students' reasoning processes*. Paper presented at the *14th International Conference on Technology in Mathematics Teaching*, Essen, Germany.

Hattie, J., & Timperley, H. (2007). The power of feedback. *Review of Educational Research*, 77(1), 81–112. doi: https://doi.org/10.3102/003465430298487.

Hughes, J., Thomas, R., & Scharber, C. (2006). *Assessing technology integration: The RAT–replacement, amplification, and transformation-framework*. In C. Crawford (Ed.), *Proceedings of Society for Information Technology & Teacher Education International Conference* (pp. 1616–1620). Chesapeake, VA: Association for the Advancement of Computing in Education (AACE).

Jerrim, J. (2016). PISA 2012: How do results for the paper and computer tests compare? *Assessment in Education: Principles, Policy & Practice*, 23(4), 495–518. doi:https://doi.org/10.1080/0969594x.2016.1147420.

Johnson, C. I., & Priest, H. A. (2014). The feedback principle in multimedia learning. In R. Mayer (Ed.), *The Cambridge Handbook of Multimedia Learning* (pp. 449–463). Cambridge: Cambridge University Press.

Kinnear, G. (2019). *Delivering an online course using STACK*. In *Contributions to the 1st International STACK Conference 2018*. Fürth, Germany: Friedrich-Alexander-Universität Erlangen-Nürnberg.

Laborde, C. (2001). Integration of technology in the design of geometry tasks with Cabri-Geometry. *International Journal of Computers for Mathematical Learning*, 6(3), 283–317. doi:https://doi.org/10.1023/a:1013309728825.

Livne, N. L., Livne, O. E., & Wight, C. A. (2007). Can automated scoring surpass hand grading of students' constructed responses and error patterns in mathematics. *MERLOT Journal of Online Learning and Teaching*, 3(3), 295–306.

Mory, E. H. (2004). Feedback research revisited. In D. Jonassen (Ed.), *Handbook of Research on Educational Communications and Technology* (pp. 745–783). New York, NY: Lawrence Erlbaum.

Narciss, S. (2008). Feedback strategies for interactive learning tasks. In J. M. Spector, M. D. Merrill, J. J. G. Van Merriënboer, & M. P. Driscoll (Eds.), *Handbook of Research on Educational Communications and Technology* (pp. 125–144). Mahwah, NJ: Erlbaum.

Narciss, S., Sosnovsky, S., Schnaubert, L., Andrès, E., Eichelmann, A., Goguadze, G., & Melis, E. (2014). Exploring feedback and student characteristics relevant for personalizing feedback strategies. *Computers & Education*, 71, 56–76. doi:https://doi.org/10.1016/j.compedu.2013.09.011.

Olsher, S., Yerushalmy, M., & Chazan, D. (2016). How might the use of technology in formative assessment support changes in mathematics teaching? *For the Learning of Mathematics*, 36(3), 11–18.

Pea, R. D. (1985). Beyond amplification: Using the computer to reorganize mental functioning. *Educational Psychologist*, 20(4), 167–182. doi:https://doi.org/10.1207/s15326985ep2004_2.

Pinkernell, G., Gulden, L., & Kalz, M. (2019). *Automated feedback at task level: Error analysis or worked out examples – which type is more effective?* Paper presented at the *14th Conference on Technology in Mathematics Education*, Essen, Germany.

Ras, E., Whitelock, D., & Kalz, M. (2015). The promise and potential of e-assessment for learning. In P. Reimann, S. Bull, M. Kickmeier-Rust, R. Vatrapu, & B. Wasson (Eds.), *Measuring and Visualizing Learning in the Information-Rich Classroom* (pp. 21–40). New York, NY: Routledge.

Ripley, M. (2009). Transformational computer-based testing. In F. Scheuermann & J. Björnsson (Eds.), *The Transition to Computer-Based Assessment* (pp. 92-98). Luxemburg: Office for Official Publications of the European Communities.

Rittle-Johnson, B., & Koedinger, K. R. (2005). Designing knowledge scaffolds to support mathematical problem solving. *Cognition and Instruction*, 23(3), 313-349. doi:https://doi.org/10.1207/s1532690xci2303_1.

Sangwin, C. (2013). *Computer Aided Assessment of Mathematics*. Oxford, UK: Oxford University Press.

Sangwin, C., & Köcher, N. (2016). Automation of mathematics examinations. *Computers & Education*, 94, 215–227. doi:https://doi.org/10.1016/j.compedu.2015.11.014.

Shute, V. J. (2008). Focus on formative feedback. *Review of Educational Research*, 78(1), 153–189. doi:https://doi.org/10.3102/0034654307313795.

Stacey, K., Price, B., Steinle, V., Chick, H., & Gvozdenko, E. (2009). *SMART assessment for learning*. Paper presented at the *Conference of the International Society for Design and Development in Education*, Cairns, Australia. Retrieved from http://www.isdde.org/isdde/cairns/pdf/papers/isdde09_stacey.pdf.

Stacey, K., Steinle, V., Price, B., & Gvozdenko, E. (2018). Specific mathematics assessments that reveal thinking: An online tool to build teachers' diagnostic competence and support teaching. In T. Leuders, J. Leuders, K. Philippou, & T. Dörfler (Eds.), *Diagnostic Competence of Mathematics Teachers – Unpacking a complex construct in teacher education and teacher practice* (pp. 241–261). New York: Springer.

Stacey, K., & Wiliam, D. (2013). Technology and assessment in mathematics. In M. A. Clements, A. J. Bishop, C. Keitel, J. Kilpatrick, & F. K. S. Leung (Eds.), *Third International Handbook of Mathematics Education* (pp. 721–751). New York, NY: Springer.

Threlfall, J., Pool, P., Homer, M., & Swinnerton, B. (2007). Implicit aspects of paper and pencil mathematics assessment that come to light through the use of the computer. *Educational Studies in Mathematics*, 66(3), 335–348. doi:https://doi.org/10.1007/s10649-006-9078-5.

Wright, D., Clark, J., & Tiplady, L. (2018). Designing for formative assessment: A toolkit for teachers. In D. R. Thompson, M. Burton, A. Cusi, & D. Wright (Eds.), *Classroom Assessment in Mathematics* (pp. 207–228). Cham, Switzerland: Springer.

6

DIGITAL PLATFORMS FOR MATHEMATICS TEACHER CURRICULUM DESIGN

Affordances and constraints

Ghislaine Gueudet,

UNIVERSITY OF BREST, FRANCE

Birgit Pepin,

EINDHOVEN UNIVERSITY OF TECHNOLOGY, THE NETHERLANDS

Scott Courtney,

KENT STATE UNIVERSITY, USA

Zeger-Jan Kock,

EINDHOVEN UNIVERSITY OF TECHNOLOGY, THE NETHERLANDS

Morten Misfeldt and

UNIVERSITY OF COPENHAGEN, DENMARK

Andreas Lindenskov Tamborg

UNIVERSITY OF COPENHAGEN, DENMARK

Introduction

Our study concerns the impact of educational technologies on mathematics teachers' work. While some studies evidence the difficult integration of Information and Communication Technology (ICT) by teachers (e.g., Bretscher, 2014), others observe the generalized availability and use of digital resources (Siedel & Stylianides, 2018). This raises the issue of the conceptualization of educational technology and/or digital curriculum resources. Pepin, Choppin, Ruthven and Sinclair (2017a) introduce the following distinction between digital curriculum resources (DCR) and other educational technologies:

> It is the attention to sequencing – of grade –, or age-level learning topics, or of content associated with a particular course of study (e.g., algebra) – so as to cover (all or part of) a curriculum specification, which differentiates DCR from other types of digital instructional tools or educational software programmes.
>
> (ibid., p. 647)

The work presented here focusses on a specific kind of digital instructional tool – namely, the digital platform. Digital platforms vary widely in terms of access, institutional demands and function. Some of them can be considered DCR (as they incorporate aspects of the curriculum); others only provide teachers with tools to design, incorporate and structure DCR.

In the next (second) section, we present selected research literature concerning digital platforms. In the third section, we present the context of our study: curriculum and digital education issues in the Netherlands, France and Denmark. Fourth, we introduce the theories framing our study. Fifth, we present and analyze one digital platform (i.e., case) from each of the three countries. In the final section, we compare and discuss the three cases.

Digital platforms: Related works

In many countries, digital platforms are now provided by educational authorities, teacher associations or other institutions (e.g., commercial or not-for-profit business, university) for use in grades K–12 and post-secondary education. As characterized by Ghazawneh and Henfridsson (2015, p. 199), digital platforms "exist in a socio-technical setting, where governance, decisions, and actions of actors related to the platform impact its evolution and prosperity". Some digital platforms are provided by teacher associations with the aim of fostering collective design and the sharing of resources (e.g., Australian Association of Mathematics Teachers; National Centre for Excellence in the Teaching of Mathematics in the United Kingdom). Other digital platforms are offered by educational authorities with the explicit aim of teacher professional development in alignment with the objectives of the national curriculum.

It can be argued that digital education platforms have become more commonplace. However, they are often denoted by terms that are vaguely defined, not quite synonymous or interchanged without meaningful definition (including, e.g., e-learning platforms, online learning environments or DCR). These platforms "are complicated mixtures of software, hardware, operations, and networks" (Kenney & Zysman, 2016, p. 64) that can support teachers with aspects of their daily work, including task selection, activity and lesson preparation; long-term lesson planning; and teacher collaboration. Some digital (education) platforms incorporate static mathematical texts, dynamic exercises and videos; offer various software packages; or allow for communication between teachers and their students, between different teachers and between teachers and other participants (e.g., parents).

In terms of frameworks for analyzing digital platforms, according to de Reuver, Sørensen and Basole (2018), most research on digital platforms adopts existing categories as their unit of analysis, predominately either the platform's technical components or organizational elements.

When digital platforms incorporate aspects of the curriculum, they are closely related to e-textbooks, defined by Pepin, Gueudet, Yerushalmy, Trouche and Chazan (2015) as

> an evolving structured set of digital resources, dedicated to teaching, initially designed by different types of authors, but open for re-design by teachers, both individually and collectively.
>
> (ibid. p. 644)

Indeed, the concepts and methods developed for the analysis of paper textbooks are still relevant for the analysis of e-textbooks (e.g., Rezat, Fan, Hattermann, Schumacher, & Wuschke, 2019). Nevertheless, we claim that other tools are needed to consider the possibilities opened by the digital means incorporated in the e-textbook. We introduced such concepts and tools in Gueudet, Pepin, Sabra, Restrepo and Trouche (2018); in this chapter, we present them in the "theory" section and use them for the study of digital platforms.

We compare and contrast digital education platforms in three countries: Denmark, France and the Netherlands. Each platform is designed specifically for use by pre-tertiary teachers and endorsed by municipal, provincial or national authorities. Our analysis heeds de Reuver et al.'s (2018) recommendation to explore alternative units of analysis that "allow for the theoretical treatment of the fundamental characteristics of digital platform characteristics and innovation arrangements" (p. 131). We discuss each platform's affordances and constraints, explicating the differences we observed and the impact in terms of teachers' curriculum design (Pepin, Gueudet, & Trouche, 2017b) as aspect/s of teachers' documentation work.

Contexts

In the following paragraphs, we briefly describe the contexts surrounding each country's digital platform. In 2008, the Dutch governmental advisory board for education published a report that encouraged the use of open educational resources (OER) and recommended support for teachers to "arrange and develop" digital resources (Onderwijsraad, 2008). In response to these recommendations, a freely accessible platform, "Wikiwijs", was launched in 2009 by the government-funded ICT service provider to the field of education ("Kennisnet"). On this platform, teachers (at any education level) can find, use, adapt, design and share digital educational resources.

In France, a new curriculum for primary and lower secondary schools (from grades 1 to 9) was introduced in 2016. This curriculum reform coincided with the so-called Digital Plan for School of the Ministry of Education. A platform, called the

"Digital Resources for School Database" (DRSB) was designed by private publishers and selected and paid for by the ministry; it opened in September 2017. Its use is not compulsory for teachers (a direct use by students is not intended). Teachers may also use the digital platform proposed by the Sésamath association (Gueudet, Pepin, Sabra, & Trouche, 2016), which covers the same mathematical content.

In 2014, the Danish government decided on a national digitalization strategy for the Danish public sector. This strategy included an initiative directed towards compulsory school called "The User Portal Initiative", which required all municipalities in Denmark to purchase and implement a digital learning platform before the end of 2017 (KL, 2014). Instead of centrally developing a national learning platform, the Ministry of Education and Local Government Denmark decided to develop 64 functional requirements for the platforms and leave their production up to private manufacturers (Misfeldt et al., 2018). The responsibility for choosing, purchasing and implementing a platform was then left to the individual municipalities based on their ongoing initiatives, aims and challenges (KL, 2014). Amongst other things, the 64 requirements specified that the platforms should provide teachers with a digital infrastructure to design teaching sequences (by adding, e.g., content, learning objectives, PDF documents, links, digital textbook material), to access digital commercially produced textbooks, to evaluate students' performances and to allow teachers to distribute teaching content to their students. The platforms were also required to include a student interface, where students could access lesson plans and the resources included therein. Five digital learning platforms met the 64 functional requirements, including the "Meebook" platform, which has been analyzed for this chapter.

In the following sections, we introduce the theoretical frame of (a) the documentational approach and (b) connectivity used in this study, and we present our research questions. Next, we present and analyze each national case. Finally, we confront the cases and draw conclusions.

Theoretical frames: The documentational approach, connectivity

The theory that frames our research is the documentational approach to didactics (DAD; Trouche, Gueudet, & Pepin, 2019). According to the DAD perspective, mathematics teachers interact with a variety of resources in their professional activity: textbooks and software but also students' productions and discussions with colleagues.

Teachers search for resources, select them, modify them, combining different kinds of resources: we call this a teacher's documentation work. During this activity (and for their professional objectives), teachers develop documents. A document is composed of the recombined resources and a scheme of use of these resources, which is defined as "a stable organisation of the activity for a given aim" (Vergnaud, 1998). The development process of a document is called documentational genesis. This process has two interrelated components: the instrumentation and instrumentalization processes. The affordances of the resources influence the schemes developed by the user: this is the instrumentation process. For example, the affordances offered by a platform to propose different exercises to different groups of students can stimulate and facilitate teachers to develop their differentiation practices. At the same time, the user adapts and transforms resources according to her or his knowledge and beliefs: this is called the instrumentalization process. For example, a teacher can select interactive exercises on a platform for her or his students and ask them to produce detailed solutions on paper beside the results produced on the computer. This is a personal appropriation of the platform typically associated with a stable organization that can be described as "for work on exercises, the student must produce written notes presenting and explaining his/her solution of these exercises".

We contend that digital education platforms "encourage" specific instrumentation and instrumentalization processes. The design tools of such platforms can support the original designs of teachers and potentially enhance their design capacity (Pepin et al., 2017a). This is directly linked with instrumentalization since these tools allow the teacher to design according to his/her pre-existing schemes. At the same time, a platform can structure and support teachers' design practices: through the mathematical content it offers, how the content can be sequenced and through particular features that are offered for the lesson designs. In a DAD perspective, this corresponds to instrumentation processes; its outcome is a modification of the teachers' schemes. Hence, we consider that it is especially important to analyze the affordances and constraints of digital platforms.

Documentation work holds a central place in teachers' work, and over time, they develop structured sets of resources, called (a teacher's) resource system. Teachers' documentation work and their resource systems are influenced by social (e.g., a community of mathematics teachers working in the same school) and institutional (e.g., local educational authorities) environments.

For investigating affordances/constraints of e-textbooks in terms of opportunities for learning, Pepin et al. (2015) introduced the concept of "connectivity", distinguishing between (1) macro-level connectivity (e.g., connections made between the book and other websites or between the resource systems of users) and (2) micro-level connectivity (e.g., internal mathematical connections made by the authors of the e-textbook between different representations, between the mathematical content and real-life contexts). We claim that the concept of connectivity, introduced for e-textbooks, is also relevant for the study of digital platforms and will illustrate it here. In this study, we use the notion of connectivity to investigate the affordances and constraints of digital platforms in three national contexts (Denmark, France, the Netherlands), and we link these to potential or actual instrumentation and instrumentalization processes of teachers using these platforms.

Hence our research questions are as follows:

(1) How can each digital education platform's connectivity be described at the micro- and macro-levels? What affordances and constraints does this connectivity offer in terms of potential instrumentation and instrumentalization processes?

(2) Looking across the national cases (i.e., Denmark, France, the Netherlands), are there any significant differences? If so, what are the potential consequences of these differences in terms of teachers' documentation work? What could be the reasons for potential differences?

Three cases of platforms: The Netherlands, France and Denmark

In this section, we successively present our analyses for each national case in order to answer the first research question presented earlier. For each case, we start with a presentation of the platform's structure. Then we discuss its connectivity at the macro-level and micro-level and potential instrumentation and instrumentalization processes for teachers using the platform.

Wikiwijs: A teacher resource platform in the Netherlands

In the Netherlands, schools and teachers are free to create/design their own set of educational resources, be they digital or traditional (i.e., print), freely provided or commercially produced, as long as they teach the content according to national curriculum frameworks. Regarding Wikiwijs[1] (see context section), it was expected that it would draw approximately nine million separate visits in 2019 (Kennisnet, 2019). However, in an online survey amongst mathematics teachers (Kock & Pepin, 2019), the use of Wikiwijs scored relatively low, compared to other digital resources (e.g., digital versions of mathematics textbooks, textbook companion websites).

Presentation of the platform structure

Wikiwijs is accessed through a browser. Its main page shows (1) a "search" function and (2) a "create or share" function. The search function can be used to search for educational resources in a set of educational databases and repositories, including Wikiwijs itself. It uses metadata to enable filtering and meaningful searches, such as the educational level, examination syllabi learning aims, types of materials (e.g., test, task, lesson), subject content, source (e.g., YouTube, Wikiwijs) and quality labels (e.g., assigned by a group of specialist teachers). The "create or share" function offers the possibility to create/design new, or modify existing, structured sets of educational resources called "arrangements" or to share files and links.

Typically, Wikiwijs arrangements contain a menu structure and content consisting of text in combination with (links to) pictures, audio/video, documents and links to websites. Authors may link to (parts of) other Wikiwijs arrangements or include a modified version of these arrangements in their own arrangement. Additional functionality enables the creation of different types of questions or quizzes with feedback. Each Wikiwijs arrangement ends with a colophon indicating, amongst others, the author(s) and a Creative Commons copyright notice. Teachers and students can access Wikiwijs arrangements through the search function or by URL and can download them (pdf or eBook). Teachers can link to, or import,

Wikiwijs arrangements into the digital learning environment of their school. This integration allows for differentiation of content and storage of students' test results.

Presently, individual teachers, groups of teachers and organizations (e.g., VO-content, see www.vo-content.nl) have created Wikiwijs arrangements.

Wikiwijs at the macro-level

The platform allows for various types of connections between Wikiwijs arrangements and other resources (macro-level connectivity): the Wikiwijs arrangements can be searched using educationally meaningful terms and used/downloaded in different formats, re-used and modified. Teachers can work on learning arrangements individually or collaboratively, and a system of peer evaluation and certification of the learning arrangements is in place. Hyperlinks can be created for students from a Wikiwijs learning arrangement to other learning arrangements and resources external to Wikiwijs (e.g., YouTube). Wikiwijs facilitates formative and, under some conditions, summative student assessment.

A macro-level constraint is that Wikiwijs offers no area for teacher-teacher or teacher-student communication, and teachers have no access to students' work or test results.

Wikiwijs at the micro-level: The topic of "similarity"

In April 2019, a search for Wikiwijs arrangements on mathematics created by (groups of) teachers obtained 350 results. We examined a selection of these results, by different authors, and selected one arrangement for further analysis. This was a Wikiwijs arrangement on "similarity" for grades 7 to 9 created by a school's mathematics teachers on the basis of a VO-content Wikiwijs arrangement. We looked for mathematically relevant connections (micro-level connectivity). Our observations were the following:

- In the Wikiwijs arrangements, we identified a few connections to previous knowledge, different concepts, disciplines or everyday situations.
- In order to connect to different students' needs, some learning arrangements used colour-coded tasks for high-achieving students. In general, the standard menu structure of the Wikiwijs arrangements appeared to favour a linear path through the content.
- Connections were made to different mathematical representations (e.g., geometric and algebraic representations) and different moments of appropriation: explanatory text, diagrams, formulas and explanatory videos. Dynamic mathematical content (e.g., the manipulation of mathematical objects in graphs) was technically not supported.
- In terms of assessment, the Wikiwijs arrangement on "similarity" is connected to paper-and-pencil student exercises, with "check your answer" buttons, and assessment in the form of a link to an external assessment rubric.

To summarize, the documentation work of the authors of the Wikiwijs arrangement on similarity was facilitated by the macro-level connectivity of the platform. It allowed the authors to connect to an open educational resource, to appropriate and to modify it to their needs. The potential for micro-level connectivity was used to some extent. Constraints of the platform (e.g., the linear content presentation and the lack of support for dynamic mathematical content) may have impacted teachers' work (the instrumentation process). We conjecture that teachers' knowledge and beliefs concerning mathematics and the curriculum also influenced how the opportunities for connectivity were used (the instrumentalization process) because the platform allowed for more types of content-related connections than we observed. However, this conjecture is a topic of further research.

The DRSB in France

As presented earlier, in France, the DRSB was launched in September 2017. In what follows, we will focus on the part of the platform dedicated to mathematics for grades 7 to 9 (called "cycle 4" since the curriculum reform).

Presentation of the platform structure

The DRSB platform for mathematics at cycle 4 is called "BAREM[2]" (meaning Bank of Resources for Mathematics). The platform offers resources of different kinds. Some resources are called "bricks" (3,056 "bricks" are available); others are called "modules" (394 modules) and associate several "bricks". Twenty kinds of "bricks" exist – for example, exercise, ICT exercise (involving software), interactive exercise, differentiated exercise, mind map, video and teacher's sheet. The teacher can find resources on BAREM using several criteria: level, kind of resource, competencies (corresponding to official competencies of the curriculum), difficulty, theme, plus any kind of freely chosen keyword. The resources can be proposed to students working on the platform or downloaded by the teacher; most resources can be modified. On BAREM, the teachers can also upload their own resources and create their own modules or "paths", combining several bricks. Teachers can share resources with colleagues from their school.

BAREM at the macro-level

The BAREM platform has clearly been designed to support the implementation by mathematics teachers of the new curriculum. The content corresponds to the new curriculum, incorporating, in particular, new content concerning algorithmics (Gueudet, Bueno-Ravel, Modeste, & Trouche, 2017). Moreover, several kinds of resources or tools correspond to the official policy priorities. The resources called "ICT exercises" support the use of GeoGebra and spreadsheets. The resources called "differentiated exercises" propose three versions of the same problem text that are more or less guided. One of the search tools concerns the official competencies. We interpret this in terms of instrumentation processes: the platform is likely to foster evolutions in the teachers' practices concerning the mathematical content taught, the use of ICT, the differentiation in class and the organization of the teaching in terms of competencies.

At the same time, the platform offers many possibilities for teacher design and instrumentalization. At the macro-level, BAREM offers possibilities of connections with the user's resource system by downloading the platform's resources and uploading their own resources on the platform, between different teachers by sharing resources and between the teachers and their students. Nevertheless, the platform does not provide the possibility to contact the authors in order to suggest modifications. Moreover, BAREM is only linked with the other platforms from the same publisher (concerning sciences and German for cycle 4), and there are no external links.

BAREM at the micro-level: Functions in cycle 4

Functions are introduced in cycle 4 in the French curriculum. In BAREM, the keyword "functions" leads to 244 bricks (and nine "modules" associating several bricks). In terms of micro-level connectivity, we observe the following:

- Connections between registers (graphic, table, algebraic). For example, 25 of the 54 proposed exercises involve changes of registers;
- Connections between different domains: 34 problems are proposed, and all of them involve other domains or real-life situations;
- Connections with software: 41 exercises are "ICT exercises" using GeoGebra, a spreadsheet or Scratch (a free programming language and online community);
- Connections between concepts. One mind map about functions explicitly represents connections between different concepts concerning functions; and
- Connections to various students' needs: a reduced amount, with only four differentiated exercises proposed.

The micro-level connectivity of BAREM is important, concerning registers and other domains, but there are no possibilities of assessment and no interactive exercises.

Digital learning platforms in Denmark

As mentioned, there are different platforms available for schools in Denmark. Here, we focus on one of the most widely used of these platforms called Meebook.[3] The Danish platforms do not include content. Rather, platforms are designed to support teachers in using a new curriculum (implemented in 2016), which focusses on learning objectives organized into competence areas, skills and knowledge. The new curriculum also introduces a prescribed workflow, where teachers are expected to plan lessons by first selecting a learning objective from the national curriculum and translating it into a more concrete, lesson-specific objective. Therefore, learning objectives are intended to function as the springboard for teachers' lesson designs. Next, we illustrate how Meebook supports this lesson development format, as well as other macro- and micro-level connections enabled by the platform.

Presentation of the platform structure

Meebook is a browser-based software that can be accessed through a computer, a tablet or a smartphone by logging on to an account. Access to the platform is paid for by the municipality. As with other available Danish platforms, Meebook supports the workflow associated with the curriculum standards by a "forløbsbygger" (teaching sequence editor). The forløbsbygger enables teachers to access learning objectives via the platform (such as "students can categorize polygons according to their side lengths and angles"[4]), supports teachers' interpretations of learning objectives and allows teachers to choose content and resources in accordance with the learning objectives for the given lessons. Danish platforms also enable teachers to assess students according to the predefined learning objectives.

Meebook at the macro-level

Regarding connectivity at the macro-level, Meebook allows teachers to "publish" (make public) their lessons and search for other teachers' published lessons based on subject and grade level. Furthermore, content from commercial content providers (e.g., textbook publishers) can be accessed through Meebook, both directly and by accessing other teachers' lessons. The platform also allows for teacher-student interactions; students can upload their work, and teachers can provide feedback. Moreover, the platform is built to support the implementation of the objective-oriented reform of the new curriculum and to increase connectivity to national curriculum standards. As in the case of France, we consider this an attempt to foster instrumentation in that the platform is designed with an aim at affecting teachers' practices. In the Danish case, this attempt concerns teachers' use of curriculum standards *and* the workflow associated with them.

Meebook at the micro-level

Meebook does not allow for micro-connections between mathematics education resources that are supplied by the platform in itself; it is not possible to analyze how a specific theme is presented on Meebook. Meebook's teaching sequence editor, however, supports connectivity amongst a number of different resources that teachers can access and integrate in the platform itself. These possibilities for micro-level connections include the following:

- *Connections between resources of different modalities*
- *Connections between different student activities*, such as tasks, evaluations, assessments and portfolios

- *Connections between resources and student needs.* Teachers can assess students' abilities prior to a lesson and assign resources that match level and need
- *Connections between different e-textbook resources.* Teachers can combine and integrate resources in the platform
- *Connections between learning objectives.* Teachers can access learning objectives from the national curriculum and assign one or more objective to a lesson or a sequence of lessons

Besides e-textbook material, none of the aforementioned resources (e.g., tasks) are provided directly by the platform. However, by integrating learning objectives from the national curriculum as a resource, the platform does provide an infrastructure for micro-level connectivity between activities, resources and representations that concern the same component(s) of the national curriculum standards.

Discussion

In this section, we answer our second research question concerning the commonalities and differences observed, their causes and potential consequences for teachers (see Table 6.1).

TABLE 6.1. Comparison of the macro-level connectivity of the three platforms

Connections with…	*Wikiwijs*	*BAREM*	*Meebook*
The official curriculum	Yes, by attaching labels to learning arrangements	Yes, through the official competencies in particular	Yes, strong connections through the objectives
External resources	Yes	No	No
The users' resource systems	Yes, possibilities to upload and download resources	Yes, possibilities to upload and download resources	Yes, possibility to upload
Between teachers	Yes, through email links and by supporting teamwork on learning arrangements	Yes, possibility to share resources with colleagues	Yes, but only with the teachers who use the same platform
Between teachers and students	No	Yes	Yes

Macro- and Micro-level connectivity

In terms of macro-level connectivity, the Dutch Wikiwijs platform is the most developed. Specifically, with its "search" function, the Dutch platform offers links to a variety of internal and external resources, in addition to potential connections between the platform and the users' resource systems. The French BAREM platform offers many potential connections between the users' resource systems and the platform, albeit no connections to external websites. The Danish Meebook platform has the least developed macro-level connectivity: no external links and restricted possibilities for sharing resources with colleagues.

In terms of micro-level (internal) connectivity, the Wikiwijs platform offers the possibility of internal connections (with the "design" function leading to "arrangements") linking content and representations, assessment and differentiation. The BAREM platform offers connections between representations, in particular, between different aspects of the same concept through mind maps but no internal assessment system. The Meebook platform offers no mathematical content but proposes tools that can foster connections between different students' activities. It incorporates an internal assessment system and offers possible connections with different students' needs.

Potential instrumentation processes

The potential instrumentation processes that we consider here relate to a possible influence of the platform on the documentation work of a teacher using it concerning all kinds of choices within this documentation work: for example, the structure of the lessons prepared with the platform (e.g., incorporating differentiation) and their content (e.g., using specific software).

The Dutch Wikiwijs platform is likely to generate instrumentation processes, leading (e.g., with the "design" function) to a linear structure of the arrangements produced by the teacher. In France, the content offered by the BAREM platform (corresponding to the new curriculum) and some of its functionalities (e.g., the search by competencies), are clearly chosen to generate instrumentation processes. The Danish Meebook platform appears to be the most oriented towards predefined (by the Danish Ministry) instrumentation processes. When teachers design a lesson on Meebook, they use a template (and their own content), which requires teachers to define the learning objective/s *before* connecting to their own content and potentially the textbook. Although lesson design grounded on learning objectives is an established best practice, many teachers base their planning of a lesson (or sequence of lessons) on "what comes next in the textbook" or the utilization of a specific activity. Therefore, Meebook-designed lessons follow precise objectives which are selected from the national standards.

Potential instrumentalization processes

All three platforms allow teachers to adapt, modify and transform resources according to their knowledge and beliefs and thus serve as tools for teacher design. As such, each platform can foster instrumentalization processes: the teachers will use the tools and/or the contents proposed by the platform according to their pre-existing schemes (Vergnaud, 1998). Teachers can even introduce their own micro-level connections, producing lessons with links between concepts or between representations—links not suggested by the platform but arising from teachers' own conceptions of teaching and learning mathematics. Nevertheless, according to our analyses of "arrangements" designed by teachers on Wikiwijs, it appeared that teachers did not fully exploit the potential micro-level connections of the platform. In this case, the instrumentalization processes appear to be associated with teachers' selection of only *some* aspects of the platform.

Understanding the differences

The observed differences seem strongly linked to national educational policies, which are reflected in the affordances and constraints of the platforms. In Denmark, the use of the platforms is compulsory, and their features are chosen to oblige teachers to design objective-driven lessons according to new national standards. In France, the digital platform was designed to support the implementation of the new curriculum. In the Netherlands, the platform was linked to a policy supporting the use of OER.

Research focussed on comparing textbooks from different countries has demonstrated that textbooks are strongly influenced by national cultures and policies (Pepin, Gueudet, & Trouche, 2013). Concerning digital platforms, we observed even stronger policy influences evidenced by different opportunities for "making connections". This variation led to different objects and processes of teacher documentation work that were difficult to compare. Indeed, such differences were much more than a matter of presenting mathematical content: on the Wikiwijs platform, part of the content is designed by users, whereas the Meebook platform does not offer such opportunities. Each platform offers tools for "making connections" between different objects and people, and these tools shape the very nature of curriculum design. Hence it can be argued that the three platforms, through their affordances and constraints (i.e., through dis/connectivity), embed the views of their respective educational authorities, not only on teachers' classroom practices but also on the documentation work of teachers out of class.

Final considerations and perspectives

We claim that studying the evolutions resulting from the use of digital platforms requires specific theoretical tools developed within the DAD: the instrumentation and instrumentalization processes and the concept of connectivity. In this chapter, we illustrated how the study of affordances and constraints of digital platforms in three national contexts can deepen our understanding of these evolutions. Our study here focussed on potential consequences of these platforms on teachers' work and naturally raises the question of the actual uses. In the case of Denmark, some observations of use already took place (e.g., Gueudet et al., 2019), and there was evidence that selected users reacted against the strong instrumentation processes resulting from the use of Meebook. In France, a questionnaire evidenced a reduced use of BAREM by mathematics teachers, but it has not been possible to organize direct observations of use. In the Netherlands, there is evidence that selected platforms (including new ones) are being used, but direct observations of the different platforms' use by mathematics teachers have not been conducted. We contend that the study presented here must be complemented by analyses of teachers' actual use of platforms, thus investigating the realization (or not) of the potential connections observed.

Notes

1 www.wikiwijs.nl.
2 https://www.barem-hatier.fr/; a detailed presentation of the platform can be found at https://www.barem-hatier.fr/article/vision-d'ensemble-des-ressources.
3 Meebook is described in detail here https://meebook.com and https://meebook.com/forside/folkeskolen/.
4 Ministry of Education, Danmarks Læningsportal: https://emu.dk/sites/default/files/2019-08/GSK%20-%20F%C3%A6lles%20M%C3%A5l%20-%20Matematik.pdf.

References

Bretscher, N. (2014). Exploring the quantitative and qualitative gap between expectation and implementation – a survey of English mathematics teachers' use of ICT. In A. Clark-Wilson, O. Robutti, & N. Sinclair (Eds.), *The Mathematics Teacher in the Digital Era* (pp. 43–70). Dordrecht: Springer.

de Reuver, M., Sørensen, C., & Basole, R. C. (2018). The digital platform: A research agenda. *Journal of Information Technology*, 33(2), 124–135.

Ghazawneh, A., & Henfridsson, O. (2015). A paradigmatic analysis of digital application marketplaces. *Journal of Information Technology*, 30(3), 198–208.

Gueudet, G., Bueno-Ravel, L., Modeste, S. & Trouche, L. (2017). Curriculum in France: A national frame in transition. In D. Thompson, M.A. Huntley, & C. Suurtamm. *International Perspectives on Mathematics Curriculum*, (pp. 41–70), Charlotte: International Age Publishing.

Gueudet, G., Pepin, B., Courtney, S., Kock, Z.-J., Misfeldt, M., & Tamborg, A. (2019). *Digital platforms and Mathematics Teachers documentation work*. In Rezat, S., Fan, L., Hattermann, M., Schumacher, J., & Wuschke, H. (Eds.). *Proceedings of the Third International Conference on Mathematics Textbook Research and Development*, (pp. 91–116). Paderborn: Universitätsbibliothek Paderborn.

Gueudet, G., Pepin, B., Sabra, H., Restrepo, A. & Trouche, L. (2018). E-textbooks and connectivity: Proposing an analytical framework. *International Journal for Science and Mathematics Education*, 16(3), 539–558.

Gueudet, G., Pepin, B., Sabra, H., & Trouche, L. (2016). Collective design of an e-textbook: Teachers' collective documentation. *Journal of Mathematics Teacher Education*, 19(2–3), 187–203.

Kenney, M., & Zysman, J. (2016). The rise of the platform economy. *Issues in Science and Technology*, 32(3), 61–69.

Kennisnet. (2019). Jaarplan 2019 [Annual plan 2019]. Retrieved from https://www.kennisnet.nl/fileadmin/kennisnet/publicatie/Jaarplan_2019.pdf.

KL (2014). Aftale om konkretisering af det fælles brugerportalsinitiativ for folkeskolen. [Agreement about the specification ofThe User Portal Initiative for compulsory schools] Retrieved from: https://www.stil.dk/-/media/filer/stil/pdf16/141015-aftaletekst-om-brugerportalinitiativet.pdf

Kock, Z.-J., & Pepin, B. (2019). *Secondary school mathematics teachers' selection and use of resources.* In U. T. Jankvist, M. Van den Heuvel-Panhuizen, & M. Veldhuis (Eds.), *Proceedings of the Eleventh Congress of the European Society for Research in Mathematics Education,* (pp. 4262–4263). Utrecht, the Netherlands: Freudenthal Group & Freudenthal Institute, Utrecht University and ERME.

Misfeldt, M., Tamborg, A. L., Qvortrup, A., Petersen, C. K., Svensson, L. Ø., Allsopp, B. B., & Dirckinck-Holmfeld, L. (2018). Implementering af læringsplatforme: Brug, værdier og samarbejde. *Læring og Medier,* 18.

Onderwijsraad. (2008). *Onderwijs en open leermiddelen* [Education and open educational materials]. The Hague, the Netherlands: de Onderwijsraad.

Pepin, B., Choppin, J., Ruthven, K., & Sinclair, N. (2017a). Digital curriculum resources in mathematics education: Foundations for change. *ZDM, The International Journal in Mathematics Education,* 49(5), 645– 661.

Pepin, B., Gueudet, G., & Trouche, L. (2013). Investigating textbooks as crucial interfaces between culture, policy and teacher curricular practice: two contrasted case studies in France and Norway. *ZDM, The International Journal on Mathematics Education,* 45(5), 685–698.

Pepin, B., Gueudet, G., & Trouche, L. (2017b). Refining teacher design capacity: Mathematics teachers' interactions with digital curriculum resources. *ZDM Mathematics Education,* 49(5), 799–812.

Pepin, B., Gueudet, B., Yerushalmy, M., Trouche, L., & Chazan, D. (2015). E-textbooks in/for teaching and learning mathematics: A disruptive and potentially transformative educational technology. In L. English & D. Kirshner (Eds.) *Handbook of Research in Mathematics Education*(3rd ed., pp. 636–661). New York: Taylor & Francis.

Rezat, S., Fan, L., Hattermann, M., Schumacher, J., & Wuschke, H. (Eds.). (2019). *Proceedings of the Third International Conference on Mathematics Textbook Research and Development.* Paderborn, Germany: Universitätsbibliothek Paderborn.

Siedel, H., & Stylianides, A. J. (2018). Teachers' selection of resources in an era of plenty: An interview study with secondary mathematics teachers in England. In L. Fan, L. Trouche, C. Qi, S. Rezat, & J. Visnovska (Eds.) *Research on Mathematics Textbooks and Teachers' Resources: Advances and Issues* (pp. 119–144). New York: Springer.

Trouche, L., Gueudet, G., & Pepin, B. (2019). *The 'Resource' Approach to Mathematics Education.* New York, NY: Springer.

Vergnaud, G. (1998). Toward a cognitive theory of practice. In A. Sierpinska & J. Kilpatrick (Eds.), *Mathematics Education as a Research Domain: A Search for Identity* (pp. 227–241). Dordrecht: Kluwer.

7

CAS FROM AN ASSESSMENT POINT OF VIEW

Challenges and potentials

Uffe Thomas Jankvist,

AARHUS UNIVERSITY, DENMARK

Jonas Dreyøe,

UNIVERSITY OF COPENHAGEN, DENMARK

Eirini Geraniou,

UNIVERSITY COLLEGE LONDON, UK

Hans-Georg Weigand and

UNIVERSITY OF WÜRZBURG GERMANY

Morten Misfeldt

UNIVERSITY OF COPENHAGEN, DENMARK

Introduction

Digital technologies (DT), such as graphical calculators, spreadsheets and other mathematical software, in particular Computer Algebra Systems (CAS) and Dynamic Geometry Environments (DGE), have become an integral part of mathematics teaching and learning during the last three decades. Not surprisingly, this also changes the usual assessment situations in the subject. DT, of course, provide a myriad of new possibilities in terms of assessment, but they also give rise to new and unforeseen challenges. Despite existing examinations of DT in many countries, the literature on this particular topic is somewhat limited (e.g., Drijvers et al., 2014).

In this chapter, we address the use of CAS in summative, as well as formative assessment situations. More specifically, we ask the question, *In what ways do CAS augment and change the assessment situation?* Our answer to this question is by example, meaning that we draw on carefully selected illustrative examples from Denmark, Germany and the United Kingdom, where CAS have entered heavily into the mathematics programmes during the past decade or so, yet in very different ways.

The theoretical approach for developing the discussion draws on constructs related to assessment in general, assessment (both summative and formative) involving CAS and the theories of instrumental genesis and instrumental orchestration (TIO). We expect – and hope – that the discussions in this chapter can form the basis for designing mathematically "meaningful" assessments involving CAS, exemplified by the situations in Denmark, Germany and the United Kingdom. By "meaningful", we refer also to the traditional goals of mathematics education being valid if we use DT.

Theoretical constructs related to assessment

Assessment as a field of research is rather large and diverse. Overall, the aim of assessment can be divided into two areas: *formative* and *summative*. The purpose of formative assessment is to assess what the students know to be able to improve the students' learning and optimize the next step in their learning (Taras, 2005). Harlen (2005) calls the formative assessment "helping for learning", and you could thus say that formative assessment is forward-looking. Summative assessment, on the other hand, is backward-looking, as the purpose of this type of assessment is to summarize what students have learnt, which at times also includes comparing the result to a standard.

Alignment in an educational context refers to the learning activities in the instructional process and the activities in the assessment situation corresponding well to each other. Hatie (2009) argues, "Any course needs to be designed so that the learning activities and assessment tasks are aligned with the learning outcomes that are intended in the course. This means that the system is consistent" (p. 6). Biggs and Tang (2011) developed the concept of constructive alignment, which includes a dialectic process: students construct meaning from the learning activities they engage in, and the teacher aligns these planned activities with the learning outcomes in the curricula. Biggs and Tang (2011) mention that students always learn what they think they will be tested on and thus refer to the *backwash effect* (or backwashing). The planning of teaching is, as mentioned, affected by two things: the learning outcomes in the curriculum and the assessment at the end of the teaching. The backwash effect describes the process of the assessment affecting the planning of the teaching.

Theoretical constructs related to the use of CAS

In any interaction with a digital tool, in particular CAS, students are expected to reach certain learning outcomes supported by the design principles of the tool itself but also the teachers' input when and if that's feasible. There is a certain process that takes place during those student interactions and that has been characterized as (the theory of) instrumental genesis (e.g., Guin & Trouche, 1999). It involves the process of transforming digital tools into mathematical instruments, which become part of students' cognitive scheme (Vergnaud, 2009) and can be used to support students' learning of mathematical concepts (Guin & Trouche, 1999; Artigue, 2002). Students need to develop the instrumental genesis and effective procedures to manipulate an artefact (Guin & Trouche, 1999). There are two processes involved, and these are referred to as the instrumentation and the instrumentalization processes. The instrumentation process focusses on the user, the user's behaviour, the utilization of his or her schemes and how these emerge and evolve when executing a certain task. We consider a scheme to be the systematic process on how to use a certain tool to achieve a certain goal. During their interactions, students may acquire knowledge that may lead to a different use of the artefact, referred to as *instrumentalization*. Instrumentalization, therefore, focusses on the artefact and involves the evolution of the use of the artefact in question, perhaps in various stages, such as a discovery phase of the artefact's functionalities, a personalization phase or even a transformation phase, where the user can add "new" functionalities to the artefact. The process of transforming digital tools into mathematical instruments is certainly a lengthy process, and instrumental genesis is expected to develop over time (Artigue, 2002). The process of instrumentation is closely connected to the digital tool serving an *epistemic* purpose (Jankvist & Misfeldt, 2015), meaning that it is used to create understanding or support learning within the user's cognitive system (Artigue, 2002; Lagrange, 2000; Trouche, 2000). By contrast, when a digital tool is to create a difference in the world external to the user, it may be said to serve a *pragmatic* purpose (Jankvist & Misfeldt, 2015). CAS, and DT in general, serve, of course, both pragmatic and epistemic purposes, but any use which is only, or mainly, pragmatic is according to Artigue (2010) of little – or even negative – educational value.

The various orchestrations described in TIO involve whole-class teaching (Drijvers et al., 2014) and are used to describe the role of the teacher in guiding students during their interactions with a digital resource and supporting them in mastering this resource usage and learning the mathematics involved. While these orchestrations are indeed used to describe and analyze a teacher's strategies when teaching mathematics with the use of DT, one could debate how they could also be linked to the assessment of students' progress. Teachers' various orchestrations are aimed at encouraging the effective use of DT in the classroom and preparing students for any final assessment with DT.

Assessing students involving the use of CAS is not as straightforward as one may think. Questions arise as to what students are assessed on. Are they assessed on their mathematical learning outcomes? Or are they assessed on their ability to use CAS in questions to solve the given mathematical tasks? Or are they assessed on both? Furthermore, do CAS serve only as a tool in the hands of the students, or are they used in the hands of the teachers (assessors) to automate the assessment of the students' performance? To answer such questions, we also discuss the concept of lever potential (Dreyfus, 1994) and the black-boxing principle (Buchberger, 1990, 2002). Technology may assist both teachers and students in handling/managing technical operations and be used as a lever potential, shifting the focus from any technical aspects to the learning of mathematics, problem solving and reasoning (Dreyfus, 1994; Winsløw, 2003). In more detail, the lever potential would support students by outsourcing certain mathematical processes and allowing them to focus on the mathematical learning objective in question. There are, however, certain challenges in such an outsourcing process. For example, students may lose sight of the solution to a mathematical activity and what it entails if certain actions are being taken care of by the digital resource. There is a certain "magic", or black boxing, when the digital resource carries out part of mathematical processes involved. Relevant research (e.g., Lagrange, 2005; Jankvist & Misfeldt, 2015; Jankvist, Misfeldt & Aguilar, 2019) has shown that students may be able to perform CAS-based mathematical activities, for example, but are still unable to understand the underlying processes. This leads to students being trained on how to use DT for certain mathematical processes, thus becoming dependent on using these DT and being unable to carry out such processes when using the pen-and-paper medium instead or any other medium outside the digital resource (Nabb, 2010).

Research review on DT in summative assessment situations

In contrast to the rather diverse, yet quite well-established, practice of using DT in everyday teaching, the use of DT in assessment situations is rather regulated. For example, Buteau, Jarvis and Lavicza (2014) show that far more teachers allow students to use CAS throughout the year than in the final exams.

When using DT in assessment, it is, of course, important to consider how to evaluate students' mathematical skills based on the test situations involving available DT. This tends to require *further specification of how to document DT* work in students' written assignments and tests (Galbraith, Pemberton, & Haines, 1996; Weigand, 2013). There has also been research into how student attitudes towards DT influence their utilization of DT. The research shows that there is a substantial variety in the utilization, and it was discovered that students' attitudes towards technical skills and experience with DT heavily influence their utilization of DT (Galbraith, Goos, Renshaw, & Geiger, 2001).

Pallack (2008) suggests that students' DT use in assessment may be strengthened if DT is used in situations where the focus is not on technical skills but on broader problem solving and modelling. This, however, is not trivial because it can be difficult to create tasks that go beyond basic skills and still are easy enough for most students to engage in, which compromises the fidelity of the evaluation in the lower grade levels (Pallack, 2008; Watson & Ohtani, 2015).

Another key problem when using DT (in general as well as in relation to assessment) is the connection between procedures and conceptual understanding since this connection can easily be detached (Jankvist & Misfeldt, 2015). In an assessment perspective, this means that it is important to assess students' ability to relate critically and reflectively to answers in connection with solving tasks (van Herwaarden & Gielen, 2002). Furthermore, Kendal and Stacey (2000) show that using DT (in this case CAS) in the daily teaching improves the understanding of graphic representations and transformation between registers, whereas more classical approaches tend to favour and develop symbolic understanding and reasoning (Kendal & Stacey, 2000).

An increased use of DT in evaluations, such as national tests, will likely result in a backwash effect on the teaching and subsequently on the textbooks and lead to an increased use of DT in the teaching (Yeap, 2010). The use of DT is shown to have a positive impact on learning in relation to project work (Shafii-Mousavi & Kochanowski, 2006), modelling (O'Callaghan, 1998) and focus on competencies (Weigand, 2013). Some results point to that integrated use of DT in teaching may lead to better test performance (Nolan & Herbert, 2015; Tsuei, 2007; Weigand, 2009). However, despite the potential benefits of using DT, the risk of black boxing persists (Chavez, White, & Hock, 2006), and it is shown to be important to integrate paper-pencil techniques with DT techniques in students' work (van Herwaarden & Gielen, 2002; Kendal & Stacey, 2000). Forster (2006) stresses that alignment is present between the utilization of technology in the assessment situation and learning situations to support students' conceptual understanding in mathematics.

Research review on DT in formative assessment situations

Formative assessment, as shared by Black and Wiliam (1998), states that formative assessment is aimed at gathering information regarding students' understanding so that the teachers can adapt their teaching and the learning activities within and between lessons to benefit students' learning. In the same sense, Suurtamm et al. (2016) "use the term formative assessment to indicate informal assessments that teachers might do as part of daily instruction as well as more formal classroom assessments used to assess the current state of students' knowledge" (p. 14).

In the following, we discuss how DT may support the formative assessment process. Looney (2010), for example, referred to some key functions of DT that can support formative assessment, such as "rapid assessment, timely and targeted feedback, interactive learning, assessment and tracking" (p. 834, as cited in Dalby and Swan, 2019). She also argued about the capacity of DT in terms of gathering and processing data on a large scale but also at speed, which makes DT a powerful resource for supporting teachers in formative assessment (Looney, 2010).

A key element in any formative assessment process is the feedback the student receives (e.g., Sadler, 1998). DT may support students, and subsequently teachers, by providing instant and constructive feedback to students and, therefore, encouraging students' independent learning in addition to, or as a replacement for, teacher-led processes (Dalby & Swan, 2019). In other words, DT can enhance formative assessment processes by providing instant and appropriate feedback to students, enabling them to have a richer and beneficial learning experience. In the field of learning analytics, there has been some progress where students receive "intelligent" feedback from the digital tool they are interacting with (e.g., Noss et al., 2012; Mavrikis et al., 2012). For example, Martinez-Maldonado (2016) reflected upon teachers' classroom orchestrations and argued about conceptualizing learning analytics as tools that can be used to support a teacher in managing, adapting, scaffolding and assessing learning activities (c.f. Prieto et al., 2015, Rodriguez-Triana et al., 2018).

However, there are different types of DT. For example, you may have tools that offer an indication for whether students' submitted answers are correct or incorrect (in which case the teacher remains an important person in the learning process) – e.g., CAS or DGE, and you may have DT, such as intelligent tutoring systems or exploratory learning environments – e.g., Carnegie Learning Cognitive Tutor Algebra I (Pane, Griffin, McCaffrey, & Karam, 2013) or Fractions Lab (Hansen et al., 2016), that provide "intelligent" feedback to challenge students' thinking (in which case the teacher becomes an additional guide and support in the learning process). Considering the latter (i.e., exploratory learning environments in particular), there have been some results carried out, and tools have been developed to support teachers' practices. For example, Gueraud et al. (2009) and Ben-Naim et al. (2008) looked at ways to share the statistics of students' interactions, such as how often did a student produce a certain kind of action, and, therefore, they were able to interpret students' behaviour. Similar digital tools have been developed and integrated into larger exploratory learning platforms, such as Metafora (Dragon et al., 2013) and MCSquared (Mavrikis & Karkalas, 2017), which are aimed at helping teachers' reflections on the use of exploratory learning environments for science and mathematics learning. Similarly, Gutierrez Rojas et al.'s (2012) work focussed on tools that would inform the teacher of students' progress and need for help in the context of computer programming labs, while Amir and Gal's (2013) work provided visualizations that could offer teachers insight into students' problem-solving strategies by reviewing their plans and their interactions over a timeline. Mavrikis et al. (2019) have shared their work on teacher assistance tools that provide real-time information about students' progress and, therefore, support their assessments of students and, subsequently, their classroom orchestration.

Finally, we also consider how crucial the role of the teacher can be in using DT effectively and appropriately for formative assessment processes. Feldman and Capobianco's (2008) study of technologically enhanced formative assessment showcases how essential it is for teachers to have good knowledge of the hardware and software, as well as a good understanding of formative assessment and related pedagogies. Similarly, Dalby and Swan (2019) argued about teachers' greatest challenge in using DT in the classroom being not the DT used but their understanding of how the DT can enhance student learning. In their study, they found out that "analysing and visually representing technologically assisted processes, highlights how teachers are often essential to the construction of effective formative processes, although their role can also become more peripheral when the technology has the capacity to perform more powerful functions" (ibid., p. 843). The functions they identified include rapid assessment, timely feedback and tracking of student learning, as well as "the accessibility of summaries of student responses for teachers, from which they can quickly identify common misconceptions, and the direct formative feedback to students provided by some interactive, adaptive software" (ibid, p. 842). But the crucial aspect is "how" teachers use this information and, therefore, how they use the DTs and their affordances as educational tools.

Examples of DT in assessment situations

In this section, we provide three illustrative examples of assessment situations/tasks involving CAS, taking into account the differences between the educational systems from which these stem (Denmark, Germany and the United Kingdom). The examples are selected in such a fashion that they both cover a wide spectrum of different educational levels (lower-secondary school, upper-secondary school and university) and so that they illustrate challenges, as well as potentials, regarding the use of DT in assessment situations. At the same time, we also differ the uses of CAS in relation to "solving" the tasks, taking into account what is permitted in the respective educational systems. Yet first, a short introduction to the different national contexts concerning the use of CAS (or DT) in assessment situations.

CAS and DT in Denmark, Germany and the United Kingdom

In Denmark, CAS have been used in exams since 2005 in upper-secondary school (the classical stream, called stx) and in lower-secondary school since 2013. Both of these exams (grade 9 and stx) are two-part exams, where one part is without any aids and the other part involves aids, including CAS and DGE. The Danish school system is in that sense quite far in implementing CAS and DGE technologies in the teaching and learning of mathematics. This implementation, however, is often left up to the individual teacher on a day-to-day basis, which can be confusing for students, not least if different teachers have very different norms related to the use of CAS in the classroom (Jankvist & Misfeldt, 2015; Jankvist, Misfeldt, & Marcussen, 2016). In short, CAS persist to be a difficult story in Denmark, in particular in upper-secondary school.

The situation in Germany concerning the educational system is quite complex. There exist 16 states in Germany and each state is responsible for the educational system in the state, especially for the curriculum and the (final) examinations. Federal agreements have to be accepted by *all* 16 states. One example is the National Standards for grade 10 (KMK, 2004) and grade 12/13 (KMK, 2012), which were established as a reaction to the Programme for International Student Assessment (PISA)-Shock in 2002. Based on the National Council of Teachers of Mathematics Standards and in line with the PISA model, a three-dimensional competence model was developed with *content guidelines*,[1] *general competencies*[2] and three levels of *cognitive activation*. In PISA, these dimensions are called "overarching ideas" (content), "competencies" (process) and "competence clusters" (cognitive activation). The 16 German states agreed to integrate these standards into the state curriculum and to orient the examinations towards these standards. Moreover, they also agreed to develop a common *pool of tasks* and to take the *technology-free part* of the final examinations out of this pool.[3] However, the second part of the final examination is still different in different states, also concerning the use of DT. In some states, only arithmetic calculators are allowed (Baden-Wuerttemberg), whereas in others, graphing calculators are mandatory (North-Rhine-Westphalia); in some states, symbolic or

CAS calculators are mandatory (Thuringia), and in yet others, students can choose whether they do the examination with arithmetic or with symbolic calculators (Bavaria, Lower Saxony). Notebooks – e.g., with GeoGebra – are in general not allowed in examinations, and there are only a few model classes and laptop schools which have special state permission. All curricula in Germany emphasize – of course – the importance of DT, and teachers are asked to use them in the classroom work, but their use is not allowed even in classroom examinations.

In the United Kingdom, DT is under-used in secondary schools and its potential is not necessarily recognized by the majority of mathematics teachers (see, for example, the Joint Mathematical Council of the United Kingdom, 2011). Even though there is a wide range of DT ready to be used, such as dynamic geometry tools and CAS, spreadsheets and programming languages, these are not widely used in secondary education. The use of DT, and, in fact, ICT, has been "advised" in the UK National mathematics curriculum for Key Stages 3 and 4 (Years 7–11, 11- to 16-year-old students), and there are a number of documents shared by the Department for Education (e.g., ICT in mathematics, 2004) in an effort to support mathematics teachers in the integration of DT in their classrooms. The use of any kind of DT relies heavily upon teachers and their confidence in using DT for teaching (e.g., Bretscher, 2014). There are teachers who use only interactive whiteboards for modelling purposes and others that use specialized DT, such as CAS or GeoGebra. But as mentioned earlier, there is no formal obligation for the teachers to use such DT, and even though their use is advised by the Department for Education (ICT in mathematics, 2004), CAS or GeoGebra use, for example, is not common in the British mathematics secondary school classrooms. Moreover, national examinations with the use of DT are not part of the compulsory secondary educational system (11- to 16-year-old students). As stated in the JMC report (September 2011), a change to the assessment of the mathematics curriculum with the use of DT is necessary. However, the problem of the wider integration of DT into secondary mathematics education remains, as a change in the assessment will not necessarily encourage the further development of the classroom practices of mathematics teachers, as many of them still have limited knowledge and experience of the use of DT for mathematical teaching and learning (JMC report, 2011). In primary and secondary education, there are online websites for setting and assessing students' homework,[4] but DT are not used for any formal national summative assessments.

Things are not that different at the university level in the United Kingdom. The use of DT, and in particular CAS, relies heavily upon the teaching staff of the university in question (e.g., Hoyles et al., 2008). The use of online platforms, such as Moodle, and the use of online courses, such as Massive Open Online Courses, are often seen in British universities, but the use of CAS is not common. A 2009 report from the Higher Education Funding Council for England clearly stated the need for enhancing learning and teaching through the use of technology (https://dera.ioe.ac.uk/140/1/09_12.pdf). In an effort to share an example of the successful integration of DT, and in particular CAS, in the mathematics education scene in the

EXAMPLE 7.1 CAS IN DANISH SECONDARY SCHOOL WRITTEN ASSESSMENT

This task stems from the ninth-grade final exam with aids (December 2016) and is named "water consumption in the shower". The written test consists of a total of six tasks with four to five subtasks in each. Task 3 contains five subtasks. We consider the first of these (Figure 7.1).

In a usual paper-and-pencil setting, this task tests if students are able to translate – or mathematize – the expression "8 minutes and 30 seconds" into 8.5 min, multiply this by 14 L/min and obtain 119 L. This translation part, simple as it may seem, is nevertheless an essential element of the task and hence the assessment. Taking into account that students have access to a variety of CAS tools, we illustrate the solution of the aforementioned task by using Wolfram Alpha. As input, Wolfram Alpha accepts the sentence: "14 L per min 8 min and 30 sec". It provides an interpretation of the input for the user to check and next reveals the result "119 L" all in one go – i.e., one push of a button. In essence, this means that the essential part of the task – i.e., the mathematizing – is outsourced to the CAS tool (see Jankvist, Misfeldt, & Aguilar, 2019). Hence this element is no longer the central aspect of the assessment. To a larger degree, it becomes a "test" of whether the students know how to feed the CAS tool with the required information.

Astrid takes a shower every morning. In the shower, she uses approx. 14 litres of water per minute.

How many litres of water does Astrid use during a shower that takes 8 minutes and 30 seconds?

FIGURE 7.1 The task reads: Astrid takes a shower every morning. In the shower, she uses approximately 14 litres of water per minute. How many litres of water does Astrid use in a shower that takes 8 minutes and 30 seconds?

United Kingdom, we focus on the use of the System for Teaching and Assessment using a Computer algebra Kernel (STACK), an online tool for formative and summative assessments used at the University of Edinburgh in Scotland.

Discussion part 1: Technology making mathematics tasks obsolete

In a sense, the previous situation/solution is a joke about technology-heavy mathematics teaching (and assessment): the teacher assigns a task to the student, who types it directly into the computer system and the system gives an answer that the student then gives to the teacher. Everyone is happy, but nobody learnt any mathematics. From an assessment point of view, this is unfortunate for several reasons. Firstly, it does not really assess relevant mathematical competencies. Instead, it tests to what extent CAS has been instrumentalized by the student. DT comes to serve as a black-box device rather than a lever potential, which in this case not only black boxes the actual mathematics of the task but also the mathematization and de-mathematization processes. Secondly, from the perspective of the backwashing effect, this solution strategy promotes a type of learning situation with technology that we know from the literature is unfortunate (e.g., Artigue, 2010). We can use the instrumental approach to describe this scheme as building only on pragmatic mediations. If a pragmatic use of CAS is all that is asked for in the assessment situation, then the backwash effect ("teaching to the test") will most likely result in a pragmatic use of CAS in the day-to-day teaching as well. This, of course, is a form of alignment, yet an unfortunate one indeed.

EXAMPLE 7.2 CAS IN GERMAN UPPER-SECONDARY SCHOOL WRITTEN ASSESSMENT

In the following, we give an example from the final examination in Bavaria from 2014.[5] The final examinations are mixed problems with on the one side only pure mathematical aspects and on the other side problems with environmental aspects (we do not talk about modelling).

The exam task reads: *[Figure 7.3] shows schematically the traffic junction Altdorf near Landshut (a city in the southern part of Germany), which connects the A92 motorway and the B299 federal road. In the shown coordinate system, a unit length corresponds to 20 m, this means the diagram represents a length of 400 m in the west-east direction. In the following, the width of the streets should be disregarded. Using the drawn coordinate system in the considered range, the A92 can be modelled by the straight line with equation $y = 0.45\,x$, the B299 by the line with equation $y = -0.5\,x$.*

a) *On the basis of the model, determine the size of the angle under which the A92 and the B299 intersect.*

If someone – coming from Munich (München) – leaves the A92 to continue on to the B299, he/she takes the southern exit, which begins in the model at point P and ends at point Q. The curve of this exit should be described in the model by a polynomial function s of grade 3 whose graph matches the points P and Q. In point P, the exit should be without any kink, and in point Q, it is perpendicular to the B299.

b) *On the basis of a system of equations, determine the coefficients of the function equation of s. Indicate for each equation which demand on the road construction is taken into account.* (Figure 7.2)

We first give a short overview of the solution of this problem. In CAS, an algebraic solution might look like what is presented in Figure 7.3. Graphical illustrations of this can be seen in Figure 7.4, both from GeoGebra and Casio ClassPad.

The values of the coefficients of *s* (Geogebra) or *y4* (Casio ClassPad) show that a hand calculation is very challenging. The use of the CAS allows you to concentrate on the construction of the system of equations and the relation between the mathematics and the real-world situation.

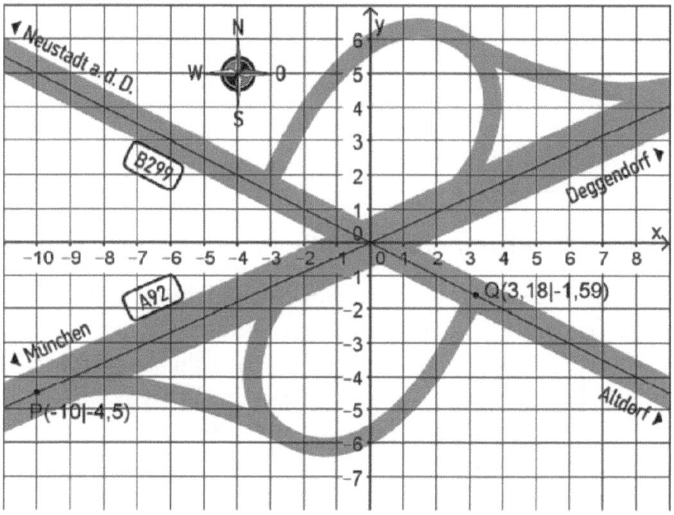

FIGURE 7.2 This is the illustration that goes along with the exam task

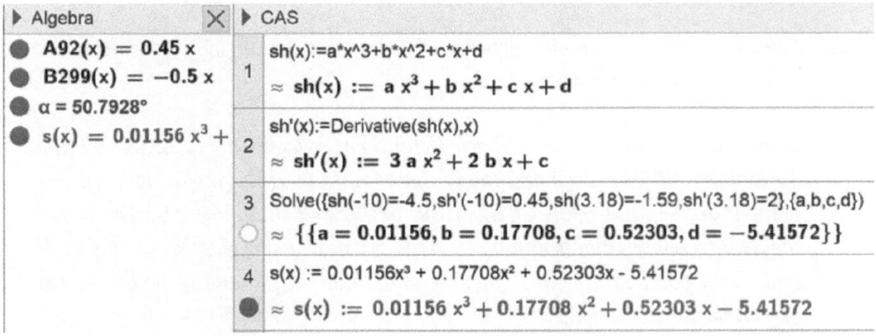

FIGURE 7.3 A CAS solution – here from GeoGebra (image created in GeoGebra geogebra.org)

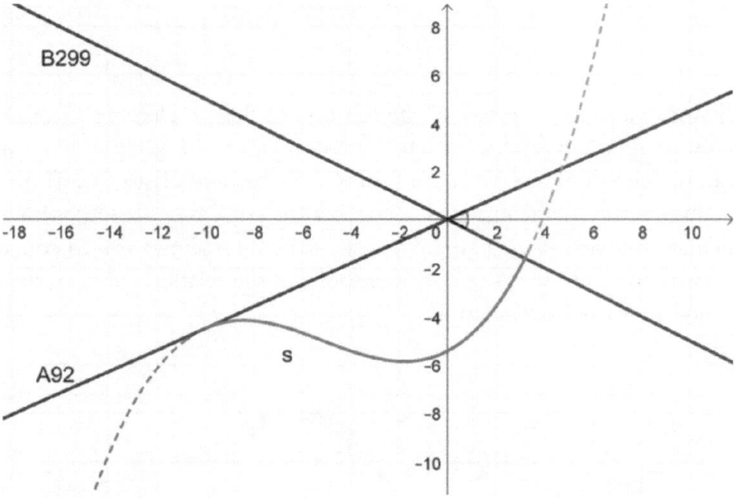

FIGURE 7.4 Graphical GeoGebra solution "s"

Discussion part 2: Making problem situations accessible with DT

In the previous example, the students are provided with the tools to do the required calculations and avoid being confused or "de-motivated" by the complex calculations and not "nice" numbers. Instead, their focus is shifted to the mathematical concepts, such as the angle created by two intersecting linear graphs and the coefficient of a cubic function 's', and therefore DT in this case is used as a lever potential. Also, GeoGebra serves an epistemic purpose by challenging the students' cognitive system by providing them with the equations and graphs of the functions in question and enabling them to identify the relationships between them. Hence these tasks to a larger extent align with an epistemic use of CAS in the classroom – unlike the previous Danish example.

Furthermore, if we look at a picture of the real traffic junction, we can see that the examination model is quite appropriate. However, it is also evident that the modelling process was already done by the authors of this problem. Rather than solving an actual problem, the "model" explores the relation between the roads, the picture of the roads and the model with graphs of functions. One could argue, of course, that the problem in many ways resembles a modelling problem. Considering the issue of alignment between the taught content and what students would be assessed on in a summative assessment, it is evident that the question does not assess students' modelling skills (representing a mathematical problem using CAS) but instead focusses on using a given model to solve the mathematical problem. Teachers reflecting on such an exam question could question their planning with regards to their teaching and the curriculum (*backwash effect*) and debate about whether the use of CAS should involve any elements of modelling "real" situations before looking at the "maths" involved. But going back to a summative assessment, such as the classical written examination, involving a modelling problem is very difficult or not possible. In a modelling process, you have to do the mathematizing, building a mathematical model, on your own. You need to have time to try different solution strategies, and you have to be allowed to follow strategies that may appear as dead-end strategies in the end. All these processes are welcome and, in fact, required in a mathematics lesson, but it would be unrealistic to be fully assessed in a written summative assessment. If you want to do a – real – modelling process in an examination situation, you have to think – beyond classical written tests – to new forms of examinations (e.g., like oral examinations or portfolios; Højgaard Jensen, 2007). Overall, this problem is a good examination problem because it is a catalyst for doing problems like this – but then as "real" modelling problems – in the classroom.

EXAMPLE 7.3 CAS IN UNIVERSITY ONLINE ASSESSMENTS IN THE UNITED KINGDOM

CAS also provide new possibilities in terms of developing digital environments that support assessment. One example of this is the tool STACK. In post-compulsory education in the United Kingdom, there has been work on the assessment of mathematics with the use of DT (e.g., Sangwin, 2013), and this is certainly a growing research field. Nevertheless, as Geraniou, Mavrikis and Margeti (2018) claimed, "Automatic formative and summative assessment has important teaching and learning implications" (p. 284), which one should consider. To display some of these implications and consider the benefits and challenges faced in the United Kingdom context, we would like to focus on an example of an online assessment system, STACK,[6] which has been used in post-compulsory education in the United Kingdom. STACK is a sophisticated assessment software for mathematics and science developed at the University of Edinburgh. The STACK system employs "the power of CAS to accurately compare mathematical expressions, automate graphical representations and perform rapid re-calculation to facilitate assessment" (Sangwin & Grove 2006 as cited in Geraniou, Mavrikis and Margeti, 2018). There are a number of university courses at the University of Edinburgh that use STACK for assessment in the form of quizzes that are given either per week or per semester and can count towards the final grade. To illustrate the STACK system, we are using an example from a quiz on the demo course titled "Calculus Refresher".

The first question, Q1, asks the student to differentiate $(2x + 2)^2$ with respect to 'x'. The student is made aware of the marking model – i.e., the question is to be marked out of 1.00 (see Figure 7.5a) – and there is always a "Check" button for the student to submit his or her answer once they finish working on it. As a student starts solving the question and types in his or her answer, the STACK's automated feedback appears beneath the student's typed answer, giving him or her hints as to why this answer would be invalid if submitted in its current state. In fact, the system's automated feedback – "This answer is invalid" – may appear until the correct syntax is used. Throughout this process, the system shows the student's answer as it interprets it and produces comments to support the student towards a correct final answer (see Figure 7.5a). Once the student clicks the "Check" button, he or she receives a green tick if the answer is correct and the message, "Correct answer, well done. Marks for this submission: 1.00/1.00" (see Figure 7.5b). If the student submits an incorrect answer, then the STACK feedback informs him or her with a red cross and a comment such as, "Incorrect answer. Marks for this submission: 0.00/1.00. This submission attracted a penalty of 0.10" (see Figure 7.5c). Such feedback can support students in monitoring their progress and being aware of their total score by the end of the quiz. In such a case, students are given the option to continue their efforts in answering a question and submitting another answer. For example, in Figure 7.5d, we can see the third attempt of the student, who corrected the power in the second bracket but also added the coefficient "$-x^4$", so the student's final answer was $3(1 - x^6)^2 \, x \, (-6x^5)$. In the second attempt, the student forgot the coefficient "6" in $(-6x^5)$. This resulted in two penalties being added and, therefore, the scoring being reduced by 0.10 twice. So the student's total score is as follows: "Correct answer, well done. Marks for this submission: 1.00/1.00. Accounting for previous tries, this gives 0.80/1.00" (Figure 7.5d).

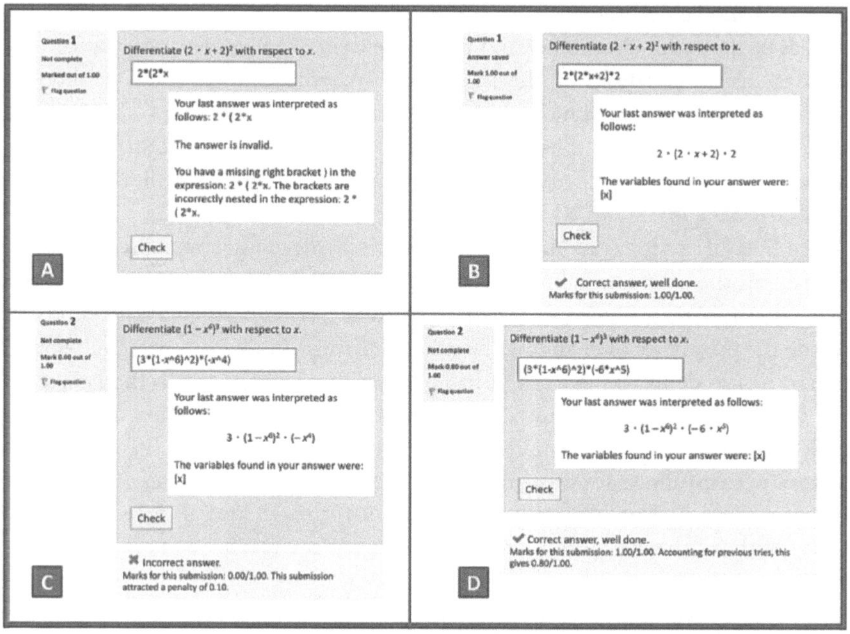

FIGURE 7.5 (a) A student's attempt to answer this question and the automated feedback from STACK, stating, "This answer is invalid" in its current state. In the automated feedback from STACK, the student sees how the answer is "translated" or interpreted by STACK. (b) The student clicked the "Check" button and the system's automated feedback states "Correct answer, well done. Marks for this submission: 1.00/1.00". (c) The student's answer in Question 2 of this quiz was given incorrectly. The automated feedback states, "Incorrect answer. Marks for this submission: 0.00/1.00. This submission attracted a penalty of 0.10". (d) A student's final answer on Question 2, following two incorrect attempts. The score is 0.80 out of 1.00, as two penalties have been applied. (All images courtesy of Chris Sangwin.)

Discussion part 3: Using DT to make examinations meaningfully efficient

Using a system like STACK, which relies on the CAS Maxima (Sangwin, 2013), is not just a matter of making better and more efficient summative assessments; tools like STACK also support entirely different approaches to formative assessment. Using STACK, students can answer questions and receive direct instant feedback, both when preparing for the teaching and in the class, which is an argument for a successful alignment in the educational context of STACK between the teaching process and both the formative and summative assessment of students. In this example, we recognize that the instant and constant feedback the STACK system provides to students promotes their reflective thinking regarding the use of formal mathematical language and promotes self-assessment processes. However, we may also raise a similar argument to that raised earlier with regards to the Danish example, that of questioning whether and to what extent CAS (as used within STACK) has been instrumentalized by the student and whether the intention of the CAS features in promoting students' reflective thinking is "put into action". So, we could still argue that in this example, there is again a risk of pragmatic mediations of this CAS, but at the same time, its use for formative assessment, as well as the instant personalized feedback, can address this issue.

The STACK system can support university teachers' instrumental orchestration by enabling them to summatively assess students while the STACK online assessment system formatively assesses them. Also, tools like STACK (and several others) provide the possibility of changing the way that formative assessment is orchestrated. Rather than students taking a test or handing in homework, which is then corrected by a teacher who provides individual feedback, STACK allows these processes to be either automated or collective. Automated feedback, of course, saves resources and thus perhaps allows for more teacher-student interactions around other matters. However, there is a risk that this change is not for the better, if, for example, the quality of the automated feedback is insufficient or the teacher time is used for less efficient purposes than individual feedback.

Collective formative assessment using a tool like STACK provides entirely new types of orchestrations. By showing typical mistakes in real time while – or immediately after – students are working in STACK allows teachers to test how their students understand a concept or react to a problem situation and then discuss the answers with the students right away in the open and collective environment of the class. Such an orchestration changes the formative assessment from being an individual dialogue between teacher and student, which due to resources will need to happen relatively seldom, to a collective – although but perhaps anonymous – class discussion of the different problems and understandings present in the class.

Conclusion: Challenges and potentials

We started this chapter by asking in what ways CAS augment and change the assessment situation in mathematics teaching. We have addressed this by looking at different examples, from Denmark, Germany and the United Kingdom. We now conclude by lining up the different influences that we have witnessed through our cases.

As seen in the three examples, the use of CAS in assessments can have different forms that meet different needs and have different challenges and possibilities attached. As illustrated through the Danish example, the use of CAS in relation to classical tasks that become more or less obsolete, or at least very easy in the context of the tool, is one type of challenge. This can be explained as the development of new instrumented problem-solving techniques, where powerful tools affect the users' scheme of the task. The backwashing effect of such an assessment situation can be rather problematic since it easily ends up promoting pragmatic strategies for solving standard tasks with CAS due to such training being aligned with the assessment. We know from the literature on the use of DT (and especially CAS) that mathematics teaching that focusses on pragmatic mediations will easily lead to learning difficulties and other challenges (e.g., Artigue, 2010). Hence such an assessment situation can negatively affect the learning outcomes of students. At the same time, removing DT and CAS from the summative assessment situation is not a solution, at least not if we want students to develop mathematical digital competencies (Geraniou & Jankvist, 2019). Hence, in unfortunate circumstances, CAS can challenge a traditional assessment approach and give rise to less valid summative assessment since we cannot be certain if we actually test the mathematical competencies of the students or if we instead test their ability to carry out simple and pragmatic computer-based solution strategies. Furthermore, this very same phenomenon might have a potentially negative effect on the teaching of mathematics because of the backwashing effect of such simple pragmatic computer-based solution schemes. In short, CAS can *challenge a traditional assessment scheme.*

A healthier strategy seems to be to develop more complex tasks (e.g., with an outset in modelling) in which the tool supports the solutions of complex tasks. Such examination tasks align with complex problem solving and project work, but it appears that this type of examination question tends to be very difficult. At least the example of an exam task from Germany is not an easy one. Yet it is an example of CAS serving as a lever potential (rather than a black box) and one which would potentially have a more positive backwashing effect from an alignment perspective. Of course, both strategies – that of the Danish and German exam tasks – do require a certain level of instrumental genesis on the students' behalf. Still, there is a world of difference between what is actually being assessed since the German exam task to a much larger extent than the Danish one actually assesses students' possession of mathematical competencies and knowledge. In that sense, the German example shows that CAS can allow for new assessment practices to evolve and that they can have some advantages over more classical approaches by allowing students to work in more realistic and complex settings. The British example displays the potential of DT in relation mainly to formative assessment and how this type of assessment with DT can support both the students and the teacher. There is value in using DT to provide teachers with summative results on students' performances (e.g., scoring in the STACK system) and skills, as well as simple descriptive statistics, such as the number of activities a student has undertaken and whether he or she has submitted correct or incorrect answers. Such information, especially when considering "in-classroom" situations, enables teachers to make informed decisions about the next steps for their students and (instrumentally) orchestrate exploratory learning that leads to successful learning outcomes (e.g., Mavrikis et al., 2019).

Answering in what ways CAS augment and change the assessment situation, we can say that our examples show that (1) CAS challenge existing assessment practices by moving them in a pragmatic direction; (2) CAS allow for new types of assessment situations to be implemented because of the increased computational power that CAS provide; finally, (3) CAS can also be used as a part of the teachers' setup of the assessment situation. By having computer algebra at hand in automated and real-time student response systems, CAS can enrich the immediate feedback that students receive.

The aforementioned insights add to the knowledge about the use of DT in mathematical assessment practices in several ways. The general tendency to use CAS less in summative assessment situations than in normal teaching (as documented by Buteau, Jarvis, & Lavicza, 2014) makes sense in reference to the Danish case. If teachers experience that using CAS in summative assessment pushes their teaching in a pragmatic direction, it is understandable that they refrain from doing so. In line with Pallack (2008), the German case shows that it is indeed possible to improve the assessment situation with CAS by focussing on broader problem-solving and modelling competencies rather than on technical skills. Issues of backwashing and lack of alignment between teaching and assessment is widely established in the literature, but the Danish case shows that the combination of CAS tools and backwashing can be especially problematic because it can drag the focus of teaching in a pragmatic direction and away from understanding the mathematics in question. The English case shows the potential of directing the computer algebra functionalities towards assessing students' solutions. In line with Noss et al. (2012), this allows students to receive immediate and intelligent feedback on their proposed solutions and further allow for new types of classroom orchestration by building both on automated feedback and providing timely information to teachers concerning the difficulties that students experience (Emin-Martinez et al., 2014).

Finally, it should be mentioned that although in this chapter we have only discussed written assessment, DT/CAS undoubtedly also have a role to play in oral examinations. In fact, it appears that many of the positive aspects of using DT/CAS for epistemic purposes in classroom situations may be more easily transferred to an oral exam situation than to a written one.

Notes

1 These are numbers, measuring, space and shape, functional connections, data and chance.
2 These are arguing mathematically; solving problems mathematically; modelling mathematically; using mathematical representations; acting mathematically on a symbolical, formal and technical level; and communicating mathematically.
3 In nearly all states in Germany, the final examinations ("Abitur") consist of two parts, a technology-free part and a second part with technologies.
4 Mathletics (http://uk.mathletics.com/), Whizz (https://www.whizz.com/), myimaths (https://www.myimaths.com/) and hegartymaths (https://hegartymaths.com/).
5 ISB: www.isb.bayern.de/gymnasium/leistungserhebungen/abiturpruefung-gymnasium/mathematik/.
6 https://www.ed.ac.uk/maths/stack.

References

Amir, O., & Gal, K. (2013). Plan recognition and visualization in exploratory learning environments. *ACM Transactions on Interactive Intelligent Systems (TiiS)*, 3(3), 1–23.

Artigue, M. (2002). Learning mathematics in a CAS environment: the genesis of a reflection about instrumentation and the dialectics between technical and conceptual work. *International Journal of Computers for Mathematical Learning*, 7(3), 245–274.

Artigue, M. (2010). The future of teaching and learning mathematics with digital technologies. In C. Hoyles & J.-B. Lagrange (Eds.), *Mathematics Education and Technology-Rethinking the Terrain. The 17th ICMI Study* (pp. 463–475). New York: Springer.

Ben-Naim, D., Marcus, N., & Bain, M. (2008). *Visualization and analysis of student interactions in an exploratory learning environment*. In the *Proceedings of the 1st International Workshop on Intelligent Support for Exploratory Environments*, Maastricht, The Netherlands. http://www.dcs.bbk.ac.uk/~sergut/isee08/.

Black, P., & Wiliam, D. (1998). Assessment and classroom learning. *Assessment in Education*, 5, 7–74.

Biggs, J. B., & Tang, C. S.-K. (2011). *Teaching for Quality Learning at University: What the Student Does* (4th ed.). Maidenhead, the United Kingdom: McGraw-Hill/Society for Research into Higher Education/Open University Press.

Bretscher, N. (2014). Exploring the Quantitative and Qualitative Gap Between Expectation and Implementation: A Survey of English Mathematics Teachers' Uses of ICT. In A. Clark-Wilson, O. Robutti, & N. Sinclair (Eds.), *The Mathematics Teacher in the Digital Era: An International Perspective on Technology Focused Professional Development* (pp. 43–70). Springer: Netherlands.

Buchberger, B. (1990). Should students learn integration rules? *SIGSAM Bulletin*, 24(1), 10–17.

Buchberger, B. (2002). Computer algebra: The end of mathematics? *SIGSAM Bulletin*, 36(1), 3–9.

Buteau, C., Jarvis, D. H., & Lavicza, Z. (2014). On the Integration of Computer Algebra Systems (CAS) by Canadian Mathematicians: Results of a National Survey. *Canadian Journal of Science, Mathematics and Technology Education*, 14(1), 35–57. doi: 10.1080/14926156.2014.874614.

Chavez, J. A., White, A. L., & Hock, C. U. (2006). *Thailand international conference on 21st century information technology in mathematics education: proceedings; September 17-20, 2006 Chiang Mai, Thailand*. Presented at *Thailand International Conference on 21st Century Information Technology in Mathematics Education*, Chiang Mai.

Dalby, D., & Swan, M. (2019). Using digital technology to enhance formative assessment in mathematics classroom. *British Journal of Educational Technology*, 50(2), 832–845.

Dragon, T., Mavrikis, M., McLaren, B., Harrer, A., Kynigos, C., Wegerif, R., & Yang, Y. (2013). Metafora: A web-based platform for learning to learn together in science and mathematics. *IEEE TLT*, 6(3), 197–207.

Emin-Martnez, V., Hansen, C., Rodrguez-Triana, M. J., Wasson, B., Mor, Y., Dascalu, M., Ferguson, R., & Pernin, J.-P. (2014). Towards teacher-led design inquiry of learning. *eLearning Papers*, 36, 3–14.

Dreyfus, T. (1994). The role of cognitive tools in mathematics education. In R. Biehler (Ed.), *Didactics of Mathematics as a Scientific Discipline*, pp. 201–211. Dordrecht, The Netherlands: Kluwer Academic Publishers.

Drijvers, P., Tacoma, S., Besamusca, A., van den Heuvel, C., Doorman, M., & Boon, P. (2014). Digital Technology and Mid-Adopting Teachers' Professional Development: A Case Study. In A. Clark-Wilson, O. Robutti, & N. Sinclair (Eds.), *The Mathematics Teacher in the Digital Era* (pp. 189–212). Dordrecht: Springer.

Feldman, A., & Capobianco, B. M. (2008). Teacher learning of technology enhanced formative assessment. *Journal of Science Education and Technology*, 17, 82–99.

Forster, P. A. (2006). Assessing technology-based approaches for teaching and learning mathematics. *International Journal of Mathematical Education in Science and Technology*, 37(2), 145–164. doi: 10.1080/00207390500285826.

Galbraith, P., Goos, M., Renshaw, P., & Geiger, V. (2001). Integrating technology in mathematics learning: What some students say. *Numeracy and Beyond*, 1(2), 225–232.

Galbraith, P., Pemberton, M., & Haines, C. (1996). *Testing to a purpose: Assessing the mathematical knowledge of entering undergraduates. Technology in Mathematics Education: Proceedings of the 19th Annual Conference of the Mathematics Education Research Group of Australasia (MERGA) June 30–July 3, 1996 at the University of Melbourne* (pp. 215–220). Melbourne [Vic.]: The Group.

Geraniou, E. & Jankvist, U. T. (2019). Towards a definition of "mathematical digital competency". *Educational Studies in Mathematics*, 102(1), 29–45.

Geraniou, E., Mavrikis, M., & Margeti, M. (2018). Digital tools for bridging the knowledge gap to university mathematics. In V.C.H. Tong, A. Standen, & M. Sotiriou (Eds.), *Connecting Research and Teaching: Students as Partners in Shaping Higher Education*. London: UCL Press.

Gueraud, V., Lejeune, A., Adam, J.-M., Dubois, M., & Mandran, N. (2009). *Supervising distant simulation-based practical work: Environment and experimentation*. In *Proceedings of ECTEL 2009*, Nice, France, (pp. 602–608).

Guin, D. & Trouche, L. (1999). The complex process of converting tools into mathematical instruments: The case of calculators. *International Journal of Computers for Mathematical Learning*, 3(3), 195–227.

Gutierrez Rojas, I., Crespo Garcia, R., & Delgado Kloos, C. (2012). *Enhancing orchestration of lab sessions by means of awareness mechanisms. EC-TEL 2012*, Saarbrücken, Germany, (pp. 113–125).

Hansen, A., Mavrikis, M., & Geraniou, E., (2016). Supporting teachers' technological pedagogical content knowledge of fractions through co-designing a virtual manipulative. *Journal of Mathematics Teacher Education*, 19(2), 205–226.

Harlen, W. (2005) Teachers' summative practices and assessment for learning – tensions and synergies. *The Curriculum Journal*, 16(2), 207–223.

Hatie, J. (2009). *Visible Learning – A Synthesis of Over 800 Meta-analyses Relating to Achievement*. London, UK: Routledge.

Hoyles, C., Forbes, I., Wolfram, C., & Leigh-Lancaster, D. (2008). Computer algebra systems in the mathematics curriculum – Report of the invitation MEI seminar supported by Texas Instruments.

Højgaard Jensen, T. (2007). Udvikling af matematisk modelleringskompetence som matematikundervisningens omdrejningspunkt – hvorfor ikke? (PhD dissertation). IMFUFA Tekster no. 458. Roskilde: Roskilde University.

Jankvist, U. T. & Misfeldt, M. (2015). CAS-induced difficulties in learning mathematics? *For the Learning of Mathematics*, 35(1), 15–20.

Jankvist, U. T., Misfeldt, M. & Marcussen, A. (2016). The didactical contract surrounding CAS when changing teachers in the classroom. *REDIMAT Journal of Research in Mathematics Education*, 5(3), 263–286.

Jankvist, U. T., Misfeldt, M. & Aguilar, M. S. (2019). What happens when CAS-procedures are objectified? – The case of "solve" and "desolve". *Educational Studies in Mathematics*, 101(1), 67–81.

Joint Mathematical Council of the United Kingdom (2011). Digital technologies and mathematics education. A report from a working group of the Joint Mathematical Council of the United Kingdom Chaired by Professor Rosamund Sutherland. Edited by Dr Alison Clark-Wilson, Professor Adrian Oldknow and Professor Rosamund Sutherland (September 2011). Retrieved in June 2019 from https://www.ncetm.org.uk/public/files/9793603/JMC_Digital_Technologies_Report_2011.pdf.

Kendal, M., & Stacey, K. (2000). *Acquiring the concept of derivative: teaching and learning with multiple representations. Conference of the International Group for the Psychology of Mathematics Education (PME 24)*, Hiroshima, Japan, (Vol. 3, pp. 127–134).

Kultusministerkonferenz (KMK) (2004). *Bildungsstandards im Fach Mathematik für den mittleren Schulabschluss*. München: Wolters Kluwer.

Kultusministerkonferenz (KMK) (2012). *Bildungsstandards im Fach Mathematik für die Allgemeine Hochschulreife*. München: Wolters Kluwer.

Lagrange, J.-B. (2000). The integration of technological instruments in education: An approach by the techniques. *Educational Studies in Mathematics*, 43(1), 1–30.

Lagrange, J. (2005). Using Symbolic Calculators to Study Mathematics: The case of tasks and techniques. In D. Guin, K. Ruthven, & L. Trouche (Eds.), *The Didactical Challenge of Symbolic Calculators Turning a Computational Device into a Mathematical Instrument* (pp. 113–135). New York: Springer.

Looney, J. (2010). Thinking deeper research paper no.1 – Part 3. Making it happen: Formative assessment and educational technologies. Promethean Educational Strategy Group. Retrieved June 2019 from http://www.21digitalclass.com/uploads/4/7/2/9/47298253/c4_-_making_it_happen_-_formative_assessment_and_education_technologies..pdf.

Martinez-Maldonado, R. (2016) *Seeing learning analytics tools as orchestration technologies: Towards supporting learning activities across physical and digital spaces. CEUR Workshop Proceedings: Proceedings of the First International Workshop on Learning Analytics across Physical and Digital Spaces co-located with 6th International Conference on Learning Analytics & Knowledge*, Edinburgh, Scotland, (pp. 70–73).

Mavrikis, M., Geraniou, E., Gutierrez Santos, S., & Poulovassilis, A. (2019). Intelligent analysis and data visualisation for teacher assistance tools: The case of exploratory learning. *British Journal of Educational Technology*, 50(6), 2920–2942. doi:10.1111/bjet.12876

Mavrikis, M., & Karkalas, S. (2017). Reflective Analytics for Interactive e-books. *IxD&A*, 33, 33–53.

Mavrikis, M., Noss, R., Hoyles, C., & Geraniou, E. (2012). Sowing the seeds of algebraic generalization: Designing epistemic affordances for an intelligent microworld. Special Issue on Knowledge Transformation, Design and Technology. *Journal of Computer Assisted Learning*. doi: 10.1111/j.1365-2729.2011.00469.x.

Nabb, K. A. (2010). *CAS as a restructuring tool in mathematics education*. In *Proceedings of the 22nd International Conference on Technology in Collegiate Mathematics*. Chicago, IL: Pearson Education Inc. Retrieved from http://archives.math.utk.edu/ICTCM/v22.html.

Nolan, C., & Herbert, S. (2015). Introducing linear functions: an alternative statistical approach. *Mathematics Education Research Journal*, 27(4), 401–421. doi: 10.1007/s13394-015-0147-x.

Noss, R., Poulovassilis, A., Geraniou, E., Gutierrez Santos, S., Hoyles, C., Kahn, K., Magoulas, G. D., & Mavrikis, M. (2012). The design of a system to support exploratory learning of algebraic generalisation. *Computers & Education*, 59(1):63–81.

O'Callaghan, B. R. (1998). Computer-Intensive Algebra and Students' Conceptual Knowledge of Functions. *Journal for Research in Mathematics Education*, 29(1), 21-40. doi: 10.2307/749716.

Pallack, A. (2008). Diagnosing and examining competences: The role of the new media. (Kompetenzen diagnostizieren und überprüfen: die Rolle der neuen Medien.). *Der Mathematikunterricht*, 54(6), 56–63.

Pane, J.F., Griffin, B.A., McCaffrey, D.F., & Karam, R. (2013). *Effectiveness of Cognitive Tutor Algebra I at Scale*. Santa Monica, CA: RAND Education.

Prieto, L. P., Dimitriadis, Y., Asensio-Pérez, J. I. & Looi, C.-K. (2015). Orchestration in learning technology research: Evaluation of a conceptual framework. *Research in Learning Technology*, 23(1). doi: http://dx.doi.org/10.3402/rlt.v23.28019.

Rodríguez-Triana, M. J., Prieto, L. P., Martínez-Monés, A., Asensio-Pérez, J. I., & Dimitriadis, Y. (2018). *The teacher in the loop: Customizing multimodal learning analytics for blended l.* In *Proceedings of the 8th International Conference on Learning Analytics and Knowledge* (pp. 417–426). New York, NY, USA: ACM. https://doi.org/10.1145/3170358.3170364.

Sadler, D. R. (1998). Formative assessment: Revisiting the territory. *Assessment in Education: Principles, Policy and Practice*, 5, 77–84.

Sangwin, C. J. (2013). *Computer Aided Assessment of Mathematics*. Oxford, UK: Oxford University Press.

Sangwin, C. J., & Grove, M. J. (2006). *STACK: Addressing the needs of the "neglected learners"*. In *Proceedings of the First WebALT Conference and Exhibition January 5–6, Technical University of Eindhoven, Netherlands* (pp. 81–95). Oy WebALT Inc, University of Helsinki, ISBN 952-99666-0-1.

Shafii-Mousavi, M., & Kochanowski, P. (2006). Service-learning projects in linked mathematics and computer technology courses. *UMAP ILAP Modules: Tools for Teaching*, 27(4), 449–468.

Suurtamm, C., Thompson, D. R., Kim, R. Y., Moreno, L.D., Sayac, N., Schukajlow, S., Silver, E., Ufer, S., & Vos, P. (2016). *Assessment in Mathematics Education – Large Scale Assessment and Classroom Assessment. ICME-13 Topical Surveys*. Switzerland: Springer.

Taras, M. (2005). Assessment – summative and formative – some theoretical reflections. *British Journal of Educational Studies*, 53(4), 466–478.

Trouche, L. (2000). The parable of the left and the pot with a spot: A study of the learning process in an environment of symbolic calculators. *Educational Studies in Mathematics*, 41(3), 239–264.

Tsuei, M. P. (2007). A web-based curriculum-based measurement system for class-wide ongoing assessment. *Journal of Computer Assisted Learning*, 24(1), 47–60. doi: 10.1111/j.1365-2729.2007.00242.x.

van Herwaarden, O. A., & Gielen, J.L.W. (2002). Linking computer algebra systems and paper-and-pencil techniques to support the teaching of mathematics. *International Journal of Computer Algebra in Mathematics Education*, 9(2), 139–154.

Vergnaud, G. (2009). The theory of conceptual fields. *Human Development*, 52(2), 83–94.

Watson, A., & Ohtani, M. (Eds.) (2015). *Task Design in Mathematics Education: ICMI Study 22*. Springer.

Weigand, H.-G. (2009). Towards a competence model for working with symbolic calculators in the frame of the function concept. *Teaching Mathematics and Its Applications*, 28(4), 196–207. doi: 10.1093/teamat/hrp022

Weigand, H.-G. (2013). *Tests and examinations in a CAS-environment – the meaning of mental, digital and paper representations. CERME 8. Proceedings of the eighth congress of the European Society of Research in Mathematics Education*, Antalya, Turkey, February 6–10, 2013 (pp. 2764–2773). Ankara; Dortmund: Middle East Technical University; ERME.

Winsløw, C. (2003). Semiotic and discursive variables in CAS-based didactical engineering. *Educational Studies in Mathematics*, 52(3), 271–288.

Yeap, B. H. (2010). Improving mathematical thinking through assessment. *Journal of Science and Mathematics Education in Southeast Asia*, 33(2), 187–197.

8

DIGITAL MAPS OF THE CONNECTIONS IN SCHOOL MATHEMATICS

Three projects to enhance teaching and learning

Martha Koch,

UNIVERSITY OF MANITOBA, CANADA

Jere Confrey,

NORTH CAROLINA STATE UNIVERSITY, USA

Alison Clark-Wilson,

UNIVERSITY COLLEGE LONDON, UK

Ellen Jameson, and

UNIVERSITY OF CAMBRIDGE, UK

Christine Suurtamm

UNIVERSITY OF OTTAWA, CANADA

Introduction

In his discussion of maps as cultural technologies, Siegert argues for a view of maps not as representations of space but as "spaces of representation" (2011, p. 14). In this chapter, we explore spaces of representation created through the digital mapping of school mathematics. We define digital mapping as the process by which information is compiled and represented as a digital image. Ifenthaler and Hanewald (2014) note that many educational settings have invested considerable resources in "digital knowledge maps", which they describe as "visual representations that enable enriching, imaginative and transformative ways for teaching and learning" (p. v). We are interested in the potential of digital maps as spaces of representation that offer a transformative view of mathematics curricula by departing from the linear structure of print documents. Digital maps can visually foreground the many connections within school mathematics and offer immediate access to related resources, such as instructional materials and current research. Given that learning outcomes for students in K–12 mathematics depend partly on how curriculum policy, design and enactment are aligned (Schmidt, Wang, & McKnight, 2005) and in light of the ways digital technology enhances the ability of communities to construct, organize and share knowledge (Stahl, 2000), we suggest that digital maps, when placed in the public domain, can facilitate the alignment identified by Schmidt et al. (2005). In doing so, digital maps can significantly enhance mathematics teaching and learning.

In this chapter, we describe three digital mapping projects: *Math Mapper 6–8*, a learning map developed in the United States; the *Dynamic Mathematics Curriculum Network*, a digital network that resulted from a Canadian research project; and the *Cambridge Mathematics Framework*, a knowledge map being developed in England. Each author describes features of their project, including **purpose**, intended **audience**, **unit size** of elements in the map, **structure and relationships**, **positionality and navigation**, processes enabling **extensibility and evolution** of the map, **language** use or mathematical register and **connections to other resources**. These brief descriptions introduce readers to the unique spaces of representation in each project. Visiting each project website will provide a more complete understanding of the features of each map.

Math Mapper 6–8 and the epistemology of learning maps (Jere Confrey)

The learning maps I have been developing are meant to provide insight into the paths a learner's developing knowledge is likely to follow as the learner moves from less to more sophisticated reasoning. The primary goal of these maps is to focus instruction around nine big ideas and to support learner-centred instruction in mathematics. I began building learning maps in 2008 when I undertook a project to visually describe the New York City mathematics curriculum standards using hexagons. In this early work, individual curriculum standards constituted the nodes for the map, and the goal was to assemble the hexagons to show users how the curriculum standards evolved across grades. This work concluded in a map of the Common Core State Standards in Mathematics (CCSS-M) for grades K–8 offering two views. One view showed a set of 18 learning trajectories (LTs). The other highlighted the grades of each set of curriculum standards (Confrey & Maloney, 2014; Confrey et al., 2011). However, the use of curriculum standards as nodes restricted the placement of any standard to a single location and because the United States curriculum standards vary in size, the map also lost consistency of scale. I created a second map.

This new map is one component of a digital learning system called Math Mapper 6–8 (MM6–8), which covers the content of middle-grade mathematics in the United States prior to a full course in algebra (Siemens & Confrey, 2015). The map can be accessed by registering an account at sudds.co. The **purpose** of the MM6–8 map is twofold. Firstly, the map creates a visual representation of the relationships amongst the big ideas and sub-constructs within middle school math. We see big ideas as concepts that anchor and coherently connect the content, processes and forms of argumentation in a discipline; they help avoid viewing mathematics as a set of fragmented topics and skills. The non-linear, hierarchical design of the map is intended to support the use of the map with a variety of curricular materials. Secondly, the map is intended to provide teachers with direct access to empirically based LTs (Clements & Sarama, 2004; Confrey, Toutkoushian, & Shah, 2019; Simon, 1995), which can guide learner-centred instruction and ground the map's related diagnostic assessments. An LT is

> a researcher-conjectured, empirically supported description of the ordered network of constructs a student encounters through instruction (i.e., activities, tasks, tools, forms of interaction and methods of evaluation), in order to move from informal ideas, through successive refinements of representation, articulation, and reflection, towards increasingly complex concepts over time.
>
> (Confrey, Maloney, Nguyen, Mojica, & Myers, 2009, p. 347)

On the map, LTs are specified for each sub-construct and connected to the curriculum standards to help assure the teacher's adequate coverage of standards while focussing their attention on student learning.

The map's **audience** is both students and teachers.[1] The map replaces the linearity of a book's table of contents in favour of multiple levels of visual illustration. The map's hierarchical structure helps to parsimoniously represent the primary relationships amongst big ideas supplemented with the magnification of additional detail. Students who use the map can see how what they are learning connects to a small but powerful set of big ideas. Teachers can use MM6–8 to reexamine instructional materials and curriculum documents, access Internet-based resources and diagnostically assess student progress along LTs (Confrey, Gianopulos, McGowan, Shah, & Belcher, 2017). The map gives teachers access to empirically established ideas about learning using LTs, and every level of the trajectories has a related set of assessment tasks. Assessment tests and practice are available to students and teachers, accompanied by intuitive student and class reports to guide diagnostically valid instructional moves.

Structure and relationships in the map are critical to its purpose. The learning map includes 9 big ideas, 25 relational learning clusters (RLCs) and 62 constructs, each of which is associated with an LT. An RLC is a set of one to four constructs that create a constellation of ideas to be learned together because their meaning derives from their relationships. For instance, in the RLC "key ratio relations" (Figure 8.1), the constructs of ratio equivalence (at the bottom), base ratio and unit ratio co-support and inform each other. Base ratio and unit ratio are parallel constructs on the map. When students move to the next RLC, "building, comparing and solving proportions," the first cluster's elements serve as a foundation.

Critical to the learning map's meaning are the underlying LTs. An LT appears when any construct in a cluster within a big idea is tapped. The LT levels are displayed from bottom to top to parallel a movement upwards in sophistication. Each LT level is labelled with a grade level to help teachers interpret what to expect from

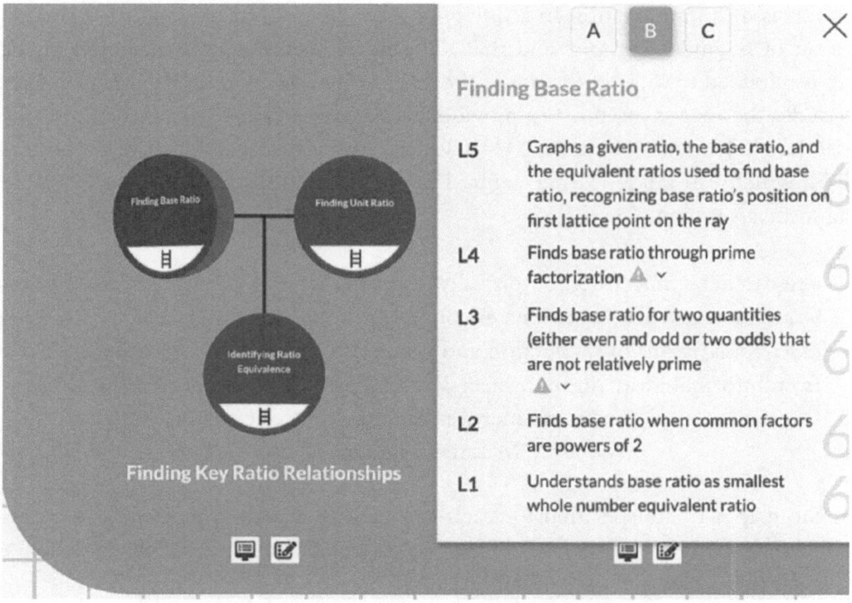

FIGURE 8.1 Learning trajectory for base ratio next to the RLC "key ratio relationships"

students across the grades. For example, the base ratio LT in Figure 8.1 has five levels. The small warning triangles signal access to common misconceptions. Tapping one reveals its description on a flip card, with the correct conception on the back.

While developing MM6–8, the **unit sizes** at each level of the map have been a constant source of challenge. Four levels (big idea → RLC → construct → LT) proved necessary to reach the level of specificity needed to capture meaningful distinctions in student thinking. Nine big ideas seemed parsimonious. Only one big idea contains more than four RLCs. To support relatively efficient, valid and reliable assessment, we limited the number of constructs in a cluster to four. We subsequently added, at the request of teachers, diagnostic assessments and practice at the construct level. Finally, the number of levels in an LT averages five, depending on the extent of detail in the research, with no LT in the map exceeding ten levels.

Positioning nodes and supporting navigation involves other design principles. The four main quadrants (number, statistics and probability, measurement and algebra) are positioned to support an overall sense of movement from bottom left to upper right. Number is an entry point, and movement is towards algebra. Likewise, RLCs within big ideas proceed left to right and bottom to top, again to signal the idea of growth. Within RLCs, lower constructs are likely to be learned before higher ones while those on the same level can be taught in any order. Navigation within the map allows easy and consistent movement with access to resources, such as curriculum standards, misconceptions and assessments handled through tabs and menus.

Coordinating topics across locations on the map was a major challenge in this project, involving significant decisions about where to locate ideas with strong connections to two big ideas. For example, linear regression is a topic with primary connections to statistics (a means to describe covarying relationships) and as an application in algebra (linear functions). For parsimony, I located topics in only one place, though I envisioned developing the idea of "wormholes" to connect disparate topics in future development.

MM6–8 is inherently **extensible and evolving**. Certain clusters have already been revised in light of subsequent research and validation of the diagnostic assessments. For instance, we added a construct, "building up/down with ratios," based on data showing students floundering in comparing ratios and finding missing values in proportions.

Mapping requires careful attention to **language**, with the challenge being to communicate to users in a precise fashion. Space is limited on a map. For instance, LT levels need to be short but convey a depth of meaning. For example, L5 of "comparing ratios" reads "compares ratios by examining relative steepness (later slope) of graph". The term, "steepness", was needed to avoid teachers prematurely introducing a formal definition of slope. Working for consistency of language use across the map has required frequent discussions and revisions based on practitioner feedback.

We built in **connections to other resources** to help users understand how the map relates to the larger educational enterprise. One such resource is access to the CCSS-M curriculum standards. Also, for each cluster, an icon allows access to illustrative curricular resources from the "resource library", which is extensible and can be filtered by construct. Another icon accesses the diagnostic assessments and practice resources.

I see the creation of learning maps as a visually expressive activity that is ongoing and dialogic. Many characteristics and design principles of MM6–8 represent the first generation of a learning map. Because our map is linked to practice most vigorously through the use of the diagnostic assessments ($n > 75{,}000$ tests), my research team studies the use of the tool, focussing on the validation of the LTs on which the maps are based (Confrey et al., 2019) and on how the data from the assessments are used to improve instruction (Confrey et al., 2018). Because our purpose is to promote learner-centred instruction, the map is subject to ongoing revision. In my mind, the admonition that "the map is not the territory" (Korzybski, 1933) reminds me that a map is fundamentally a cultural tool (Siegert, 2011) designed to help one navigate within a space. Its effectiveness is ultimately tied to how well it proves able to do that.

A participatory dynamic curriculum network (Martha Koch and Christine Suurtamm)

The Dynamic Mathematics Curriculum (DMC) Network (www.dynamicmath-curriculum.ca) emerged from our research on making connections within school mathematics more visible through the use of digital technology (Koch, Suurtamm,

Lazarus, & Masterson, 2018). The concept of the Network originated at a meeting of the Canadian Mathematics Education Study Group as we participated in a working group on reconceptualizing curricula (Davis & Kubota-Zarivnij, 2014). We began to envision a curriculum with elements of school mathematics represented as layers of connected nodes. Since then, with support from the Social Sciences and Humanities Research Council of Canada, we developed a method for engaging mathematics educators in co-creating such a curriculum network, and designed a prototype based on their contributions.

We view mathematics as emergent and deeply connected and value non-linear views of mathematics teaching and learning. We also see curriculum as inherently dynamic; a curriculum emerges as teachers transform their written curriculum standards and available resources into classroom activity. We see participatory approaches to curriculum design as ways to enhance teaching (Clandinin & Connelly, 1988; Cochran-Smith & Lytle, 2009) and recognize that mathematics teachers, working together, can create a curriculum that responds to their students' needs (Breyfogle, McDuffie, & Wohlhuter, 2010; Drake & Sherin, 2006). Researchers have also shown that mathematics learning is enhanced as teachers emphasize connections between concepts, processes and representations (Johanning, 2010). Although we find that many curricula align with this research by encouraging teachers to make connections as they teach, we also notice that curriculum documents and resources typically present mathematics as discrete strands and provide little support for teachers to make connections.

Accordingly, in Phase One of our research, we devised a method of using collaborative problem solving to prompt educators to articulate the connections they make as they engage in mathematical thinking. In a series of video-recorded sessions, we asked participants to note aspects of mathematics that came to mind as they worked on a mathematical task in a small group. Each group then created a physical model connecting the mathematical content and processes they had identified. Figure 8.2 shows one model where participants represented the mathematical processes of problem solving, collaboration and visualization using coloured pipe cleaners entwined around each of the connections the group had made between the concepts they had identified.

Analysis of these models became the basis for the first iteration of the DMC Network posted on the project website. Participants sometimes referred to their province's curriculum as they created their models, but the DMC Network itself is not linked to one set of curriculum standards. In Phase Two, we invited other mathematics educators to visit the website and provide feedback on the first iteration of the Network using an "Add to the Network" tab. After analyzing their proposed contributions, we added several new nodes and connections and renamed some existing nodes. Educators from both phases were also invited to comment on features of the website and propose relevant instructional resources. Mathematicians, teacher educators, graduate students, teachers and university-based researchers participated, and the most recent iteration of the Network shown on the website is based on their collective views and experiences.

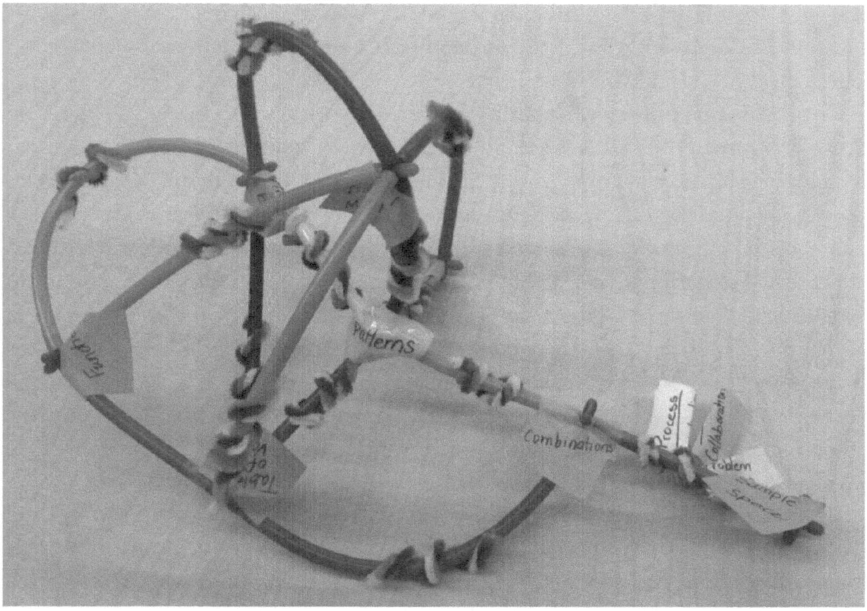

FIGURE 8.2 Research participants' model of mathematics concepts and processes

Our goal in this project has been to foreground connections in school mathematics by gathering and representing the connections educators make as they engage in problem solving. The **purpose** of the Network is to provide a prototype of what such a curriculum might look like. We anticipated that the main **audience** for the Network would be teachers as they enact their curriculum and make instructional plans. Teacher educators may also use the Network to help pre-service and in-service teachers become more aware of these connections.

As with the other maps in this chapter, we wrestled with **unit size** as we developed the Network. Participants were comfortable including different grain sizes of mathematics ideas as they built their models, and we were committed to ensuring that the Network reflected their contributions. This has resulted in a lack of consistency of scale, a concern raised in relation to MM6–8. For instance, some nodes in the Network encompass more complex concepts (e.g., "proportionality") than others (e.g., "table of values"). We also found unit size to be challenging because participants represented mathematics concepts and processes as interwoven in their models. To reflect this view, yet show some distinction between concepts and processes, we opted to use green nodes for concepts (e.g., "algebraic expressions") and blue nodes for processes (e.g., "conjecturing and verifying"). We view this as an interim approach to resolving this aspect of unit size.

Language selection and use was also a concern in our research. As we compiled the original models and reviewed Phase Two participant input, we found a range of terms used to express ideas. We created a chart of similar and related terms and referred to that chart, sometimes making additions to it, as we decided which terms to use in the Network. Our goal has been to use terms that will make sense to as many users and contributors as possible.

The **structure and relationships** in the Network reflect complexity thinking (Davis & Simmt, 2003; Doll, 2008; St. Julien, 2005). A complex system, represented as a network, is more flexible than a linear or hierarchical structure since one can move from node to node in non-linear ways. It is also generative and adaptive as a result of interactions between its elements. These characteristics, within the digital interface we designed, mean an unlimited number of nodes and connections are possible and that some nodes may connect to many parts of the Network (e.g., "algebraic expressions" currently connects to 12 nodes), while others have fewer connections (e.g., "proportionality" currently connects to 4 nodes). As it grows, the Network also reflects multiple and emergent approaches to mathematics. Indeed, the first iteration had 10 nodes with 31 connections while the iteration after Phase Two had 13 nodes with 59 connections.

With respect to **positionality and navigation**, in the DMC Network, the position of a node is determined solely by the connections between that node and other nodes. Users can click and drag any node in the network to a different place on their screen, and the connections from the node will be maintained. Clicking on a node also highlights all connections between that node and other nodes. These features, along with a +/− zoom tool, help users focus on a specific part of the Network. Clicking on a node or a connection also reveals definitions, explanations, examples, **connections to other resources** and a description of the continuum of the concept across grades. These resources are also derived from our analysis of input from participants. Readers can explore other navigational features, including some still in the design phase, by visiting our website.

As with the other maps in this chapter, **extensibility and evolution** are key to our participatory approach. The problem-solving task we gave participants in Phase One of our research tended to prompt algebraic thinking and mathematics ideas often encountered in grades 6 to 10. Engaging participants with a different task would be one way to extend the Network to other topics. In theory, the extensibility and evolution of the Network are without limits, as users contribute their ideas through the "Add to the Network" feature. That said, moderating participants' suggestions has been challenging. Thus far, the research team reviews each submission and applies guidelines to determine which proposed changes should be made in the Network. One criterion we use for nodes is trying to avoid redundancy. For connections, a key criterion is our ability to follow and verify the mathematical thinking of the proposed connection. For this reason, the "Add to the Network" tool asks participants to explain their thinking for each proposed change. Moderating these contributions is a rich research opportunity but is also time-consuming. More feasible strategies would be needed if the Network were to move from research project to widespread use.

Our project has provided insights into the ways collaborative problem-solving prompts mathematical thinking and helps teachers articulate the connections they see

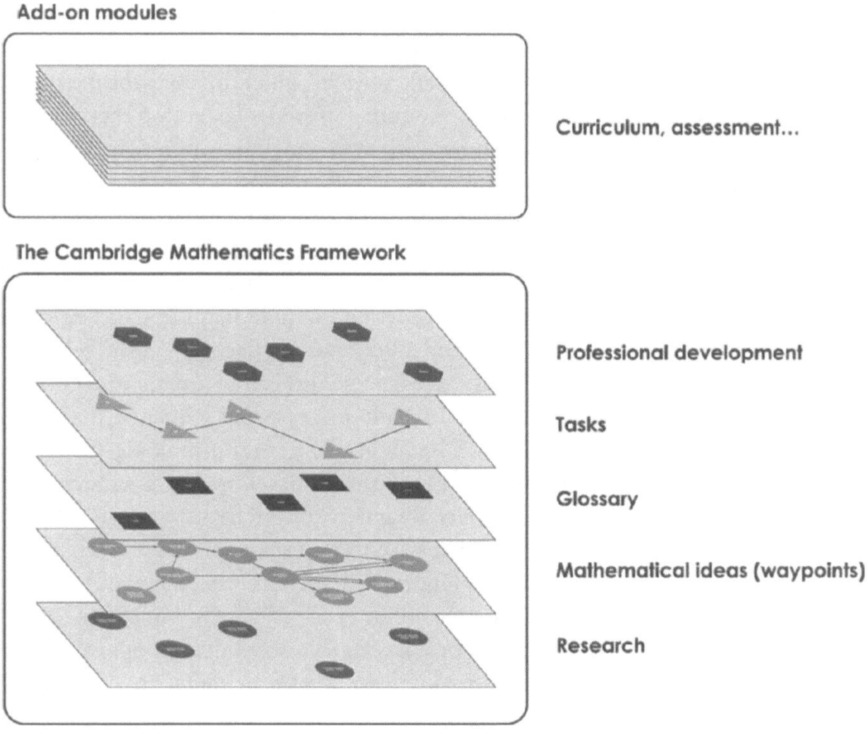

FIGURE 8.3 Connected layers within the CM Framework and external add-on modules

in school mathematics. We have been encouraged by the enthusiasm and thoughtful contributions of participants and have learned a great deal about digital affordances and constraints. Our original vision for the Network included interconnected layers of nodes, strikingly similar to the layers of the Cambridge Mathematics Framework (see Figure 8.3). We felt these layers could effectively represent the many dimensions of mathematics teaching and learning. However, some of our participants suggested a virtual reality interface would be preferable, as it would allow users to move freely within the connections space. Translating these ideas into reality captured our imagination but exceeded our financial resources. Thus some of our participants' most innovative suggestions remain future visions.

Mapping knowledge about learning in the Cambridge Mathematics Framework (Ellen Jameson)

In the Cambridge Mathematics (CM) Framework project (www.cambridgemaths.org), a team of designers, teachers and researchers are developing a framework for expressing mathematics learning experiences and key relationships between those experiences. The CM Framework is a tool for the dynamic generation of maps which highlight connections between ideas and experiences in school mathematics. These connections do not represent the paths that students follow or that teachers plan. Rather, they highlight some important ways in which mathematical experiences can contribute to or depend on one another as students develop their understanding. The maps, and associated content, are representations of knowledge about mathematics learning, as interpreted from reports of research and practice according to our methodology (Jameson, 2019; Jameson, McClure, & Gould, 2018). The maps are derived from a database which stores all connected content and a front-end interface that allows us to create, edit, search, filter, group and map the content.

Our goal in developing the ability to generate multiple maps from a larger set of connections is to provide flexibility and additional perspectives on the relationships between mathematical ideas and experiences. This can be useful when developing curriculum objectives, planning curriculum scope and sequence, designing resources and developing assessments. The ability to generate multiple maps can also contribute to teacher education and professional development. Research suggests that when people such as teachers, designers, researchers and policymakers interpret the curriculum they are working with using a shared frame of reference, it may be possible to better explain and support the coherent implementation of that curriculum (Remillard & Heck, 2014; Schmidt et al., 2005; Cunningham, 2017). Greater coherence, in turn, can have a positive impact on students' opportunity to learn mathematics by providing a shared logic for what is being taught and lowering barriers to learning that some students may otherwise not have the support to overcome (Schmidt et al., 2005).

Coherence in a mathematics curriculum depends on what can be known about mathematics and the learning process (Cobb, 1988; Tall, 2013; Thurston, 1990), as well as on what allows expert perspectives to converge (Pring, 2012; Schmidt et al., 2005) or to coordinate (Hall, Morley, & Chen, 2005; Robutti et al., 2016; Thurston, 1990) in line with some shared understanding. In our view, coherence in students' mathematical experiences depends in part on coherence in how professionals in different areas of mathematics education understand and support these experiences in their curricular environments (McClure, 2015). To address this need, the CM Framework is a network of mathematical ideas and experiences, as understood and documented by professionals in mathematics education research and practice – that is, of professional knowledge about student learning.

The **purpose** of this network is to support coherence in mathematics education. The CM Framework relates mathematical ideas to one another in a way that is not specific to any particular curriculum. At the same time, the framework relates mathematical ideas to things that may be applicable to a particular context, such as curriculum standards, tasks, assessment components and professional development activities. The framework can generate maps to represent these relationships in ways which are useful for different audiences.

Our **audiences** include anyone who holds and develops professional knowledge in mathematics education. The audiences we feel we need to provide the most support are those who work most directly with large amounts of curriculum content: curriculum designers, resource designers, teachers and teacher educators. An important part of professional knowledge in many of these roles is knowledge about someone else's knowledge. For example, maps of professional mathematicians' knowledge could be expected to be quite different from maps of teachers' or educational researchers' knowledge of student learning, which in turn would be different from maps of students' LTs. We have made decisions about how to represent researchers' and teachers' knowledge of student learning in the CM Framework based on our purpose and audiences.

After some exploratory writing, we formalized the **structure and relationships** of the CM Framework in an *ontology* which defines what features appear in the network, what they mean to us and our audiences and how they can be related to each other (Jameson et al., 2019). This ontology structures our database, determining how we create, store, search, filter and display content. The ontology helps us maintain consistency in the way mathematical ideas and experiences are represented. Similarly, it provides a guide to interpreting maps from the CM Framework for reviewers, and a version of it may help our audiences do the same.

We express mathematical content in our maps as *waypoints*, which we define as "places where learners acquire knowledge, familiarity or expertise". Our definition is based on characterizations of learning sequences by Michener (1978) and Swan (2014). Each waypoint contains a summary of the mathematical idea (the "what") and its part in the wider narrative (the "why") and lists examples of "student actions" that would provide opportunities to experience mathematics in meaningful ways. Some waypoints have particular roles; *exploratory waypoints* "indicate a place where ideas can be played with in a less formal or more playful way, as part of building mathematical intuition", and *landmark waypoints* are places where "ideas are brought together such that the whole experience may seem greater than the sum of its parts" (Jameson et al., 2019, p. 4). Waypoints are related to one another by *themes*. A theme might be a concept, skill or procedure since a student might use or develop a mathematical idea in various ways. Themes representing development have a direction, leading from one waypoint to another. Waypoints are arranged from left to right roughly according to the "order" created by these directed connections.

Research summaries are documents in the CM Framework which tell the story of a group of waypoints and themes. They include a literature review, an interactive map of the waypoints and themes and a section which describes how research has influenced the structure and contents of the map. An example of a research summary is available on our website, along with a description of our ontology (Jameson et al., 2019).

In the CM Framework, there is not one overall **unit size.** Rather, the unit size depends on the way the CM Framework is being used. For example, a single node in the network, such as a waypoint, can function as a unit, but so can a research summary and a group of research summaries could outline a subdomain. Features of waypoints, themes, research summaries and other layers (examples are shown in Figure 8.3) are designed to provide different ways of engaging with and interpreting the CM Framework for different audiences. Some people may be looking for a "way down" to get a perspective on more detailed content, while others may need a "way up" to position their detailed understanding within a bigger picture. Likewise, some may be working at a time scale of a few weeks, while others may be designing for learning over a few months, a few years or a decade.

Connections to other resources are managed within this layered structure. These resources might be for designers (such as curriculum statements for curriculum comparison or revision), for teachers and teacher educators (such as professional development activities) or for both (such as glossary definitions of mathematical terms). The structure of the CM Framework allows us to maintain connections within and between layers so that research sources can be linked to waypoints and research summaries; tasks can be linked to mathematical content, which is linked to research; and so on. This structure also opens up other possibilities for connections to outside resources depending on future directions.

In the maps that are generated from the CM Framework, positioning of waypoints is key to **navigation.** We use a left-to-right order of "dependencies," where, roughly speaking, elements towards the right have more waypoints "leading in" to them. The vertical axis does not represent anything in a typical waypoint map. However, for some purposes, we might split a map horizontally into two regions, with the upper region representing some "grouping" category, such as curriculum statements, so that connections to the network of waypoints "underneath" can be examined more easily. At other times, we use the vertical axis to stack waypoints with equivalent "dependencies" in order to demonstrate that designers have leeway when ordering these waypoints in materials they are developing. Navigation of the wider network involves building a subset of waypoints which can be added to or subtracted from the rest of the network to focus on a particular topic.

As with the other projects in this chapter, **language** is essential for articulating meaning in the CM Framework. Our research sources, which influence the content and structure of the framework, are mainly in English. However, we collaborate with researchers familiar with research in other languages (including Spanish, French, German, Japanese and Chinese) in an effort to identify perspectives and findings of which we might otherwise have been unaware – though this still leaves important gaps that we seek to fill over time. While our familiarity with English-language research strongly influences our work, we hope through collaboration that the CM Framework will be applicable outside that context as well. We also use a glossary to define key mathematical terms in the framework. The structure of this glossary allows us to offer multiple definitions, which is important since contexts for learning which draw on a particular term can vary widely. We imagine this being useful in the future as a way to create definitions specific to certain user roles or regions. In fact, we have created a survey app which allows teachers to rate a family of definitions for a term according to usefulness and accuracy for their role and context, and as of this writing, we are collecting data on our first set of terms.

The CM Framework is a large undertaking, and **extensibility and evolution** are key in our design. Not only will new areas of content continue to be added and existing areas fleshed out but also as new research becomes available, we will need to revise existing content. Our graph database, network structure and tools for online access work together to make these processes part of the work we do on a daily basis. As our work progresses, we have been piloting the use of the CM Framework for curriculum and resource design. The feedback we have received thus far has been that the flexibility and perspectives afforded by the Framework are very valuable. However, we are also mindful of what we call the "map of gaps" – the ideas and paths not shown because they are unknown to us, research not consulted because of constraints of time or language or perspectives not taken because they are too distant from our own or those of our collaborators. While any interpretation would be susceptible to these issues, maps may be more susceptible as they can "feel" visually complete. We hope the flexibility and possibility for extension and curation in the CM Framework will help audiences probe beyond one view and learn from multiple perspectives, as designing this structure has certainly done for our learning.

Affordances, challenges and future directions

These projects illustrate many aspects of the design and use of digital maps that can enhance mathematics teaching and learning. In each project, innovative approaches are used to visually represent connections within school mathematics. The connections in the CM Framework are derived from the research literature and interpreted and prioritized by a team of designers with educator backgrounds. In MM6–8, connections are based on empirically researched LTs. In the DMC Network, connections are continually being gathered from mathematics educators. Deeper mathematics understandings can be achieved for students and teachers when greater emphasis is placed on connections amongst concepts and processes (Johanning, 2010). Showing connections visually and in a non-linear manner is one of several ways these maps can contribute to more effective teaching and learning. Given the capacity of digital technology for storing and accessing information, these maps also provide immediate access to related instructional resources, curriculum standards, assessments and, in some cases, LTs or other research literature. Moreover, these maps are enormously flexible, allowing individuals to use them in ways that meet their specific needs. The connections shown in each map are more general than a single curriculum document, and each map includes tools to enable users to focus on particular subsets of the connections. In addition, in all three projects, the maps allow for multiple paths through each space.

Given these characteristics, these maps are spaces of representation that foster shared and emergent understandings of mathematics and mathematics teaching and learning. For instance, in MM6–8 teachers are able to link big ideas in mathematics to empirically based LTs, while students can see how what they are learning connects to those big ideas. Feedback from users of the CM Framework suggest it has enhanced the design of digital mathematics resources being developed by industry. Previously, such resources were often developed with more limited input from subject domain experts. In the DMC Network project, analysis of participants' comments suggests that articulating the connections they perceive and explaining those connections to others led to new ways of thinking about mathematics concepts and processes. In a similar way, the maps fostered emergent understandings across the three project teams. Recognizing commonalities in the design and development approaches we used prompted generative conversations about mathematics and led to a deeper appreciation of mathematics education in each of our jurisdictions. In these ways, whether used by policymakers, educational resource designers, researchers, teacher educators, teachers or students, digital maps can make a substantial contribution to more effective mathematics teaching and learning.

The challenges encountered in each project also suggest some future directions for this work. The unit size within digital maps is one area for further consideration. In MM6–8, consistent unit size is maintained within one level of the map, while magnification enables more fine-grained information to be displayed. In essence, multiple unit sizes are supported by the affordances of digital technology. Allowing for multiple unit sizes also creates maps that can be used in more flexible ways, as posited by the CM Framework team. Multiple unit sizes might also more closely reflect the ways users see connections in school mathematics, as noted in the DMC Network. At the same time, allowing for multiple unit sizes can compromise consistency of scale. The potential impact on users of this sort of compromise needs to be more fully understood.

Another challenge is that significant resources are required both for the initial development and the continued evolution of these maps. In each project, additional aspects of school mathematics can be included, and the connections represented can grow and change in response to new research (for MM6–8 and CM Frameworks) or as members of the mathematics education community contribute their ideas (for DMC Network). At the same time, fully representing the inherently multifaceted nature of mathematics and of mathematics teaching and learning remains somewhat elusive. The connections in school mathematics are more layered and complex than we can effectively represent, even with the affordances of digital technology. Thus Confrey looks for a wormhole to more fully connect aspects of the MM6–8 learning map, participants in Koch and Suurtamm's project suggest the use of virtual reality or the addition of sound and motion to the DMC Network and CM Framework designers endeavour to enable switching rapidly between multiple representations and views. Continuing to push the horizons towards these possibilities could mean these digital maps become even more imaginative spaces of representation, more fully capturing the richness of mathematics and the complexity of mathematics teaching and learning.

At the same time, further study of how the current iterations of these first-generation maps are being used to facilitate decision making in discussions amongst and between curriculum and resource designers, teachers, teacher educators and/or policymakers will provide important insights into how these maps contribute to more effective and equitable mathematics teaching and learning within and across educational contexts.

Note

1 MM6–8 has been developed in ongoing partnership with three districts using the map and assessments and benefitted substantially from their suggestions and observations of their practices.

References

Breyfogle, M. L., McDuffie, A., & Wohlhuter, K. A. (2010). Developing curricular reasoning for grades pre-K–12 mathematics instruction. In B. J. Reys & R. E. Reys (Eds.), *Mathematics Curriculum: Issues, Trends, and Future Directions* (pp. 307–320). Reston, VA: NCTM.

Clandinin, D. J., & Connelly, M. (1988). *Teachers as Curriculum Planners: Narratives of Experience*. New York: Teachers College Press.

Clements, D. H., & Sarama, J. (2004). Learning trajectories in mathematics education. *Mathematical Thinking and Learning*, 6(2), 81–89.

Cobb, P. (1988). The tension between theories of learning and instruction in mathematics education. *Educational Psychologist*, 23(2), 87–103.

Cochran-Smith, M., & Lytle, S. L. (2009). *Inquiry as Stance: Practitioner Research for the Next Generation*. New York: Teachers College Press.

Confrey, J., Gianopulos, G., McGowan, W., Shah, M., & Belcher, M. (2017). Scaffolding learner-centered curricular coherence using learning maps and diagnostic assessments designed around mathematics learning trajectories. *ZDM Mathematics Education*, 49(5), 717–734.

Confrey, J., & Maloney, A. P. (2014). Linking standards and learning trajectories: Boundary objects and representations. In J. Confrey, A. P. Maloney, & K. H. Nguyen (Eds.), *Learning Over Time: Learning Trajectories in Mathematics Education* (pp. 125–160). Charlotte: Information Age.

Confrey, J., Maloney, A. P., Belcher, M., McGowan, W., Hennessey, M., & Shah, M. (2018). The concept of an agile curriculum as applied to a middle school mathematics digital learning system. *International Journal of Educational Research*, 92, 158–172.

Confrey, J., Maloney, A. P., Nguyen, K. H., Mojica, G., & Myers, M. (2009). *Equipartitioning/splitting as a foundation of rational number reasoning using learning trajectories*. Paper presented at the *33rd Conference of the International Group for the Psychology of Mathematics Education*, Thessaloniki, Greece.

Confrey, J., Nguyen, K., Lee, K. S., Panorkou, N., Corley, A., Avineri, T., & … Maloney, A. (2011). Turnonccmath.Net: Learning trajectories for the k-8 common core math standards. from https://turnonccmath.net/.

Confrey, J., Toutkoushian, E. P., & Shah, M. P. (2019). A validation argument from soup to nuts: Assessing progress on learning trajectories for middle school mathematics. *Applied Measurement in Education*, 32(1), 23–42.

Cunningham, M. (2017). *Method for Developing an International Curriculum and Assessment Framework for Mathematics* (pp. 1–15) [Internal Report]. Retrieved from http://inprogressreflections.ibe-unesco.org/wp-content/uploads/2018/01/Math-Content-Reference-List-and-Coding-Scheme_Methods-Paper.pdf.

Davis, B., & Kubota-Zarivnij, K. (2014). Mathematics curriculum re-conceptualisation. In S. Oesterle & D. Allan (Eds.), *Canadian Mathematics Education Study Group Proceedings 2013* (pp. 127–137). Burnaby, BC: CMESG.

Davis, B., & Simmt, E. (2003). Understanding learning systems: Mathematics education and complexity science. *Journal for Research in Mathematics Education*, 14(2), 137–167.

Doll, W. E. (2008). Complexity and the culture of curriculum. *Educational Philosophy and Theory*, 40(1), 190–212.

Drake, C., & Sherin, M. G. (2006). Practicing change: Curriculum adaptation and teacher narrative in the context of mathematics education reform. *Curriculum Inquiry*, 36(2), 153–187.

Hall, S., Morley, D., & Chen, K. H. (Eds.). (2005). *Stuart Hall: Critical Dialogues in Cultural Studies*. London: Routledge.

Ifenthaler, D. & Hanewald, R. (Eds.). (2014). *Digital Knowledge Maps in Education: Technology Enhanced Support for Teachers and Learners*. New York: Springer-Verlag.

Jameson, E. (2019). Methodology: Building the research base (p. 21). Retrieved from https://www.cambridgemaths.org/Images/methodology-building-the-research.pdf.

Jameson, E., Horsman, R., Macey, D., Gould, T., Rushton, N., Rycroft-Smith, L., Majewska, D., Stevens, B., & McClure, L. (2019). *Ontology: Structure and meaning in the Cambridge Mathematics Framework*. Cambridge, UK: Cambridge Mathematics.

Jameson, E., McClure, L., & Gould, T. (2018). *Shared perspectives on research in curriculum reform: Designing the Cambridge Mathematics Framework. ICMI Study 24 Conference Proceedings* (p. 8). Tsukuba, Japan: ICMI.

Johanning, D. I. (2010). Designing curricula to expand and extend mathematical knowledge. In B. J. Reys & R. E. Reys (Eds.), *Mathematics Curriculum: Issues, Trends, and Future Directions* (pp. 171–180). Reston, VA: National Council of Teachers of Mathematics.

Koch, M.J., Suurtamm, C. Lazarus, J., & Masterson, L. (2018). *Making connections: Launching a co-created digital mathematics curriculum network*. In H. Weigand et al. (Eds.), *Proceedings of 5th European Society for Research in Mathematics Education Topic Conference: Mathematics Education in the Digital Age* (pp. 147–154). University of Copenhagen.

Korzybski, A. (1933). *A Non-Aristotelian Systeme and the Necessity for Rigor in Mathematics and Physics* (pp. 747–761). New York: Institute of General Semantics.

McClure, L. (2015). A Manifesto for Cambridge Mathematics. Retrieved from https://www.cambridgemaths.org/Images/cambridge-mathematics-manifesto.pdfhttps://www.cambridgemaths.org/Images/cambridge-mathematics-manifesto.pdf.

Michener, E. R. (1978). Understanding mathematics. *Cognitive Science*, 2(4), 361–383.

Pring, R. (2012). Culture: A neglected concept. In J. Elliot & N. Norris (Eds.), *Curriculum, Pedagogy and Educational Research: The Work of Lawrence Stenhouse* (pp. 49–60). London and New York: Routledge.

Remillard, J. T., & Heck, D. J. (2014). Conceptualizing the curriculum enactment process in mathematics education. *ZDM*, 46(5), 705–718.

Robutti, O., Cusi, A., Clark-Wilson, A., Jaworski, B., Chapman, O., Esteley, C., ... Joubert, M. (2016). ICME international survey on teachers working and learning through collaboration. *ZDM*, 48(5), 651–690.

Schmidt, W. H., Wang, H. C., & McKnight, C. C. (2005). Curriculum coherence: An examination of US mathematics and science content standards from an international perspective. *Journal of Curriculum Studies*, 37(5), 525–559.

Siegert, B. (2011). The map IS the territory. *Radical Philosophy*, 169, 13–16.

Siemens, G. S., & Confrey, J. (2015). Some possible implications of data-intensive research in education—the value of learning maps and evidence-centered design of assessment to educational data mining. In C. Dede (Ed.), *Data-Intensive Research in Education: Current Work and Next Steps. Report on Two National Science Foundation-Sponsored Computing Research Association Workshops* (pp. 79–87). Washington, DC: Computing Research Association.

Simon, M. A. (1995). Reconstructing mathematics pedagogy from a constructivist perspective. *Journal for Research in Mathematics Education*, 26(2), 114–145.

St. Julien, J. (2005). Complexity: Developing a more useful analytic for education. In W. E. Doll Jr., M. J. Fleener, D. Trueit & J. St. Julien (Eds.), *Chaos, Complexity, Curriculum, and Culture* (pp. 101–116). New York: Peter Lang.

Stahl, G. (2000). Collaborative information environments to support knowledge construction by communities. *AI & Society*, 14, 71–97.

Swan, M. (2014). Designing tasks and lessons that develop conceptual understanding, strategic competence and critical awareness. In J. Brocardo, A.M. Boavida, D. Delgado, E. Santos, F. Mendes, J. Duarte, M. Baia, & M. Figueiredo (Eds.), *Tarefas matemáticas: Livro de Atas do Encontro de Investigação em Educação Matemática* (pp. 9–28). Sesimbra, Portugal: Sociedade

Portuguesa de Investigação em Educação Matemática. Retrieved from https://dspace.uevora.pt/rdpc/bitstream/10174/13646/1/ATAS-EIEM2014.pdf#page=21.

Tall, D. (2013). *How Humans Learn to Think Mathematically: Exploring the Three Worlds of Mathematics*. Cambridge: Cambridge University Press.

Thurston, W. P. (1990). Mathematical Education. *Notices of the AMS*, 37(7), 844–850.

9

THE ROLE OF TECHNOLOGY IN THE PRAGMATIC-ABSTRACT CONTINUUM IN MATHEMATICS CURRICULUM DEVELOPMENT AND TASK DESIGN

Allen Yuk Lun Leung and

HONG KONG BAPTIST UNIVERSITY, CHINA

Ana Donevska-Todorova

GOETHE-UNIVERSITY FRANKFURT, GERMANY

Relevance of the theme and overview

There has always been an epistemological tension between the abstract conceptual nature of mathematics and the real-life mathematical practices. At the two extremes, a Platonist would insist that mathematical ideas have their own existence in an ideal world that is somewhat disconnected from the real world where we live, whilst for a social constructivist, mathematical ideas are socially negotiated objects used for meaningful communication about explaining real-life phenomena. These two ontological views about mathematics often create a pedagogical gap in the mathematics classroom. Platonic pedagogy favours a deductive orientation that stresses an absolute mathematical truth independent of human existence, whilst the constructivist pedagogical view takes on an inductive approach favouring realistic human experiences. These two mathematical world views, respectively abstract and pragmatic relatively, seem to be quite different and often result in a knowledge gap that prevents the crossing between two potentially compatible epistemological domains. To cross this gap, one needs a mediating object between the two worlds. Technological tools can play such a pedagogical mediating role.

This chapter begins with an explanation of what the mediating object could be. It then proceeds with explanations about the metaphor "pragmatic-to-abstract continuum" by looking at recent research. It reaches its peak by suggesting a *Dual Pragmatic-Abstract Continuum Model.* The model is then illustrated by task designs at different levels of education.

Boundary objects and epistemological continuum

A mediating object sits at the boundary between different worlds. Star and Griesemer (1989) coined the term *boundary object*, which acts as an agent to stimulate communication between different parties. Boundary objects are

> objects which are both plastic enough to adapt to local needs and the constraints of several parties employing them, yet robust enough to maintain a common identity across sites. They are weakly structured in common use, and become strongly structured in individual-site use.
>
> (Star & Griesemer, 1989, p. 393)

Technological tools possess a virtual real nature that could play the role of a boundary object to bridge the pragmatic-abstract gap.

> Digital tool has the potential to create "virtually real" objects and representations that could open up multi-interpretations of mathematical knowledge. In particular, it can act as a regulatory balance between the pragmatic and epistemic dimensions in mathematical knowledge acquisition: learning the mathematics while using the digital tool.
>
> (Leung, 2017, p. 8)

For example, Dynamic Geometry Environment (DGE) can be used as a tool for Euclidian construction of geometric figures with intact prescribed properties for the pragmatic perspective. For the abstract perspective, it can be regarded as a visualization tool to verify a geometrical theorem. A figure in a DGE can simultaneously be a drawing and a figural concept; its realization depends on the intention of the learner. Thus DGE fits the boundary object's plastic and robust characteristics. Somewhere in between, DGE's drag mode creates a continuum (sometimes literally when a DGE figure transforms under dragging) between construction, conjecture making and theorem verification (Leung et al., 2013). Hence it is possible that a digital tool can create a "pragmatic-abstract continuum" where tool-based discursive mathematical practices gradually evolve or transform into formal mathematical knowledge. This is an epistemological continuum where dialect occurs between a digital tool's pragmatic action scheme and the intended abstract mathematical knowledge to be learnt. A digital tool can be instrumented progressively to play different epistemic modes at different points in the continuum. It may start with establishing routine practices, which enable critical discernment leading to the development of situated abstraction. In this connection, Leung (2017) proposed a nested sequence of epistemic modes to guide tool-based mathematics task design consisting of three nested modes in a cognitive expansion order: establishing practices, critical discernment and establishing situated discourses. This nested sequence may demarcate different signposts in a knowledge acquisition sequence whereby, with the use of an appropriately chosen tool, learning happens as a continuum process in which routine algorithmic actions evolve or transform into internalized cognitive activities.

The pragmatic-abstract continuum relates to an idea proposed by Leung and Bolite-Frant (2015):

> Evolution of a tool's instrumented action scheme transforms the instrumental/pragmatic-semiotic/epistemic dialectic. The instrumental genesis and tool of semiotic mediation frameworks both deal with this transformation from physical tool to psychological tool. Task design needs to sequence tool-based activities to progress through the epistemological continuum where mathematical skills/techniques are gradually transformed into abstract concepts and understanding. Tools can be assigned or designed to play different epistemic roles at different points of the continuum.
>
> (Leung & Bolite-Frant, 2015, p. 220)

Instrumental genesis is an epistemic process of transforming an artefact into a user-interpreted instrument by means of the evolution of a utilization scheme for the artefact (Vérillon & Rabardel, 1995; Vérillon, 2000). Semiotic mediation (Bartolini Bussi & Mariotti, 2008) is rooted within the Vygotskian social construction of the knowledge perspective highlighting the mediating role of an artefact. Students experience the developmental process of mathematical concepts through an artefact that carries mathematical meanings as a tool of semiotic mediation. These two theories are about epistemological transformation processes of the tool and cognition.

Likewise, Donevska-Todorova (2018, p. 270) suggested a nested model of three modes of thinking and a description of concepts: algebraic, geometric and abstract-structural. The algebraic and the geometric modes refer to tangible definitions of mathematical concepts, whilst the abstract-structural mode refers to the axiomatic and formal-structural way of defining concepts through properties. The axiomatic definitions often describe a concept without explicitly stating what the concept exactly is. To illustrate this, we may consider natural numbers and counting/ordering even at the preschool level of education. One may think of Peano's axioms, which define natural numbers by five postulates without stating exactly what a natural number is. Similarly, at the secondary or university level of education, a vector may be defined geometrically as an equivalence class of objects that have the same length, direction and orientation arrows or algebraically as an ordered *n*-tuple or formally as an element of a vector space having eight axiomatic properties. Tangible and formal concepts appear as contents in curricula across all school levels (e.g., symmetric groups of plane geometric figures). Instead of considering these two types of concepts as *distinct representations* of a mathematical concept, it would be more epistemologically sound to explore a continuous connection between them.

> Abstract concepts gain their meaning in contexts and both the algebraic and the geometric modes allow such concretizations.
>
> (Donevska-Todorova, 2018, p. 269)

This idea about concretization occurring through practices whilst using the algebraic or the geometric modes is connected to Leung's (2017) nested modes in the cognitive expansion. The critical discernment happens when the learner notices the connection between the different modes and attempts to translate one into the other. By doing so, situated discourse can be established through traditional and digital tools that make such connections visible to the users.

The pragmatic-to-abstract continuum

Crossing the pragmatic-abstract gap between the two mathematical worlds is a transformative process. That is, during the crossing, a mathematical idea undergoes an evolutionary process either where a concrete physical experience gradually becomes a decontextualized concept or where a robust abstract concept is realized in a contextual real-world situation. Thus a mathematical idea can be regarded as an "idea continuum", which simultaneously possesses characteristics of the pragmatic and the abstract mathematical worlds, and at a different point on the continuum, it "collapses" into a specific situated mathematical representation. What generates this continuum? The continuum could be created by pedagogical tools that have knowledge potentials of both mathematical worlds. These tools are plastic enough for either world to use as acceptable representations of mathematical ideas. For example, a pair of compasses can be used practically to draw and to measure on the one hand, whilst on the other hand, they are conceptual tools used to represent a circle. The intention and design of the tool's usage for a mathematical idea determine the "pragmatic weight" and "abstract weight" of the tool, hence, consequently creating a tool-dependent idea continuum. Throughout the continuum, the tool goes through a pedagogical transformation (in the sense of instrumental genesis and semiotic mediation) when the pragmatic and abstract weights change. In this way, the continuum can be thought of as created by transitory "boundary objects in transformation", which can be used to map the pedagogical or curricular development of mathematical ideas.

In his research on the cognitive development of children, Jerome Bruner (1966) proposed three modes of representation:

1. *Enactive (action-based):* Representation of knowledge through actions and tool manipulation
2. *Iconic (image-based):* Visual summarization of experiences through images and diagrams
3. *Symbolic (language-based):* Use of words and symbols to describe and represent experiences

A wide body of research has been developed based on Bruner's Enactive, Iconic and Symbolic (EIS) representations, even though there are criticisms in considering the EIS as three distinct stages. In another vein, there is Tall's (2004) three worlds of mathematics: conceptual-embodied world, perceptual world and formal-axiomatic world. A procept is the amalgam of a process which produces a mathematical object and a symbol which is used to represent either process or object. Both Bruner's representations and Tall's worlds of mathematics that afford different levels of cognition (Donevska-Todorova, 2017) suggest that there might be a progressive transformation of knowledge representations starting from pragmatic experience to abstract conceptualization, or vice versa. In the teaching and learning of school mathematics, this knowledge transformation can be realized in the interplays between tool-based, graphic-based and symbolic-based pedagogical approaches. Figure 9.1 depicts a Dual Pragmatic-Abstract Continuum Model designed to explore this knowledge interplay.

The Dual Pragmatic-Abstract Continuum is represented schematically by four concentric epistemic disks (a disk is a circle together with its interior) *expanding outward from the innermost disk*. Each disk contains the previous one, suggesting a progressive accumulation and "representation" of previous practices.

- *Pragmatic-concrete disk:* Observation is done/constructed and interpreted by concrete means (e.g., physical gesture, visual perception, tool usage, measurement, counting) to represent primitive/abstract ideas.
- *Concrete-diagrammatic disk:* Concrete experiences or structured diagrams are "sketched" as intuitive images.
- *Diagrammatic-formal disk:* Primitive sketches or symbolic representation are replaced by structured formal diagrams.
- *Formal-abstract disk:* Structured diagrams are translated into formal symbolic representations or primitive symbolic representations are interpreted diagrammatically.

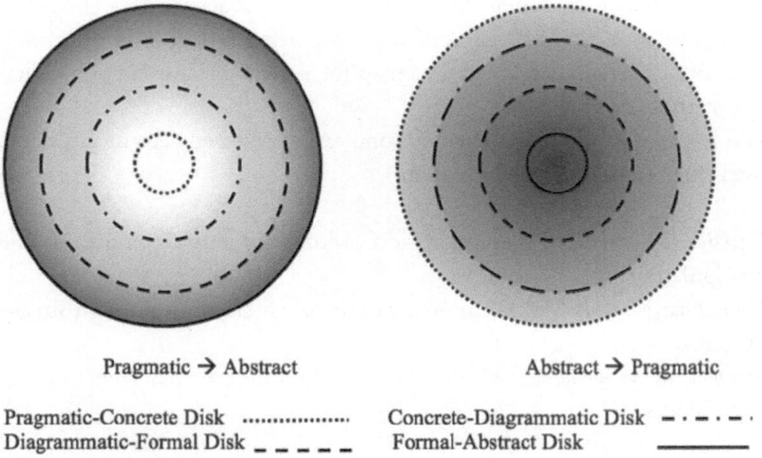

FIGURE 9.1. A dual pragmatic-abstract continuum

The pragmatic-abstract continuum is a dual concept in the sense that it is a bi-directional nested epistemic process. For example, at the outer abstract disk of the pragmatic-to-abstract direction, concrete experience, physical/virtual means, tools and diagrams still take parts in the epistemic process acting as foundational elements to build abstract symbolic understanding. Whilst at the outer pragmatic disk of the abstract-to-pragmatic direction, construction of concrete mathematical experiences is based on formal-abstract concepts. The ideal is that there are no discontinuities in the continuum if the continuum is regarded as a "natural" epistemological process where knowledge organization and complexity transform in seamless fashions. Transition going from one type of representation to another is where critical discernments occur. One can start at any point on the continuum and design or develop a pedagogical path, not necessarily linear, that is suitable to achieve a particular curriculum goal.

Task and curriculum design

With respect to the Dual Pragmatic-Abstract Continuum, mathematics task and curriculum design could gear towards designing student-directed pedagogical processes to experience mathematical ideas/concepts under technology-enhanced environments (including digital tools, digital curriculum materials, resources and tasks) focussing on interactivity, connectivity and iterative feedback. These processes should address the Dual Pragmatic-Abstract Continuum whereby tool-based discursive mathematical practices and "formal" (open to multiple interpretations) mathematical practices are mutually evolved/transformed into each other.

Symmetry of geometric objects

We now illustrate the Dual Pragmatic-Abstract Continuum shown in Figure 9.1 with the aid of specific task designs and resources utilizing traditional and technology-enhanced learning environments appropriate for three different levels of education: university, secondary school and primary school. As an example, we use symmetry of geometric objects as a mathematical content that extends across the different curricula. Thinking about the concept of symmetry from a higher point of view, it seems that at a certain point on the continuum, it splits into specific situated mathematical representations, which we discuss in detail next.

University level: The symmetry group of a geometric object is the group of all transformations under which the object is invariant under a group operation. Such transformation is an invertible mapping of the ambient space that takes the object to itself, which preserves certain structures of the object. For example, the dihedral group of degree 3 D_3 consists of rotations by 0, $2\pi/3$ and $4\pi/3$ and three reflections, and the dihedral group of degree 4 D_4 consists of four rotations by multiples of $\pi/2$ and four reflections.

Secondary school level: Utilization of DGE, such as Geometer's Sketchpad, Cabri Geometry or GeoGebra gives spatial-graphical meaning to symmetry and demonstrates the "collapse" of the concept of symmetry into its specific situated geometrical representation. For example, a GeoGebra Book[1] found in GeoGebraTube consists of six GeoGebra files about the *dihedral group* D_n (the group of symmetries of a regular n-gon) for $n = 3, 4, 5$ and 6. Pragmatic dragging of the slider shows the angle of rotations. Despite the restricted nature of the visual representation – e.g., limited to n = 3 and n = 4 – the task design has the potential to suggest an intuitive mode of description about rotations of regular n-gon. This means that a formal-structural mode of the *dihedral group* D_n is concretized for two exact figures pragmatically. Thinking further about the dragging modalities in the designed DGE, diverse gestures may apply, depending on the digital tool being used (e.g., gestures related to mouse dragging or single- or multi-touch). The repertoire of the gestures may vary whilst examining the congruence of the two dynamic regular polygons under rotation. This consideration meets design principles for technology-rich task designs regarding the reduction of the cognitive load that is taken over the technological tool (Donevska-Todorova & Weigand, 2018).

Furthermore, usage of both virtual and tangible manipulatives designed with DGE and 3D print allows for the development of different solving strategies of problems related to symmetric geometric figures, such as tessellations (Figure 9.2a) and puzzle games (Figure 9.2b) with different levels of abstraction (Donevska-Todorova, 2020). The task design may significantly differ despite usage of the same geometrical figures for constructing more complex symmetric figures by the changing of context or didactical material.

Primary school level: Representation and gesture for the concept of symmetry may appear different from those noted earlier. For a comparison, we may think of rotating the equilateral triangle (or the square) made of paper around its centre of rotation, which is kept fixed with a pushpin. In this case, traditional materials may be more profitable to enable the enactive and gestural approach related to the pragmatic-concrete disk in Figure 9.1.

FIGURE 9.2 Symmetric geometric figures created with DGE and 3D print

How does the pragmatic-to-abstract continuum for the concept of symmetry progressively develop throughout the curriculum at different levels of education? The following suggests a sequence of transitionary tasks.

Pragmatic-concrete task

Chan, Leung and Ong (2017) discussed a symmetry task using physical manipulatives in a primary school lesson where students were given small square or triangular plastic pieces to construct symmetric figures. Teachers designed a transparency toolkit consisting of Blu Tack; pushpins; transparent plastic sheets; plastic geometric shapes, such as squares and triangles; a pen; and a Styrofoam board. With the aid of a transparency toolkit, students were asked to verify whether their constructed figures have rotational symmetry or not. The discrepancy potential (Leung & Bolite-Frant, 2015), which was the degree of deviation away from the concept of symmetry resulted from the feedback from the use of the toolkit, opened up a pedagogical space for the teachers and students to discuss the critical features of a symmetric figure. This is an example of a pragmatic-concrete task on symmetry at the primary school level.

Concrete-diagrammatic task

Figure 9.3 depicts two GeoGebra files[2] that the first author used in a geometry course for in-service mathematics teachers. The intended object of learning was the symmetry of the square.

By dragging the circular dialler and the dotted mirror line, students could rotate or reflect the "top" square freely and continuously to find positions where the top and bottom (fixed) squares overlap. This dynamic visual digital tool was used to locate the eight symmetries of the square. A DGE like GeoGebra can be regarded

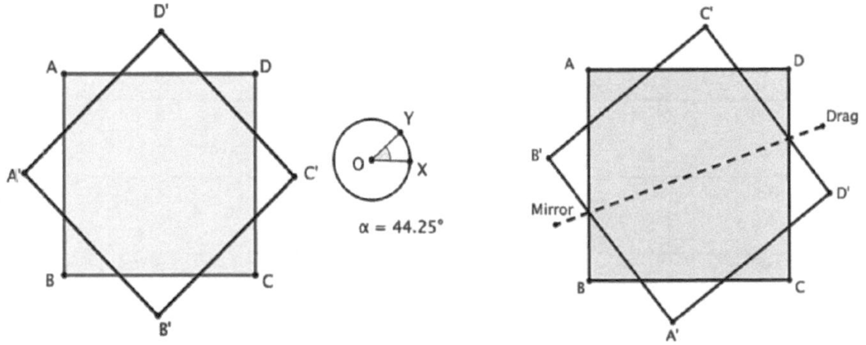

FIGURE 9.3 Snapshots of the symmetry of the square GeoGebra files (geogebra.org)

as a concrete-diagrammatic tool as it possesses the physicality of user manipulation (dragging) and spatial-graphical representation of mathematical objects. This concrete-diagrammatic task can be used at the lower-secondary school level to continue further discussion on the symmetry of the square leading to the awareness of the relationship between rotation, reflection and the "compositions" of them.

Diagrammatic-formal task

The previously described task is now accompanied by a worksheet that teaches students how to compose rotation and reflection. For example, Figure 9.4 is a transition table (a kind of boundary object) where students learn to transform what they see in the dynamic GeoGebra files to the concept of formal two-line symbols.

A similar table is in the worksheet for the square's reflectional symmetries (M1, M2, M3, M4), and an example of how to compose two-line symbols is presented. Students are then asked to complete the "multiplication/composition table" for square symmetries (Figure 9.5).

Students need to complete this "multiplication table" for square symmetries after they have learnt how to compose the square symmetries using two-line symbols. This is a diagrammatic-formal task suitable for upper-secondary students that introduces a bridging between geometry and arithmetic/algebra.

Formal-abstract task

The completed "multiplication table" (Figure 9.6) is studied under the formal-abstract language of group theory. From the table, students discover essential elements and characteristics of a symmetry group and its cyclic subgroup. This formal-abstract task is in the context of a university group theory course.

Rotation	Degree (Increasing counterclockwise)	Position Change	Two-line Symbol A=1 B=2 C=3 D=4
R1	0	A→A B→B C→C D→D	$\begin{pmatrix} 1 & 2 & 3 & 4 \\ 1 & 2 & 3 & 4 \end{pmatrix}$
R2			$\begin{pmatrix} 1 & 2 & 3 & 4 \\ \square & \square & \square & \square \end{pmatrix}$
R3		A→C B→D C→A D→B	$\begin{pmatrix} 1 & 2 & 3 & 4 \\ 3 & 4 & 1 & 2 \end{pmatrix}$
R4	270		$\begin{pmatrix} 1 & 2 & 3 & 4 \\ \square & \square & \square & \square \end{pmatrix}$

FIGURE 9.4 A worksheet for students to learn to transform what they see in the dynamic GeoGebra files to the concept of formal two-line symbols (geogebra.org)

o	R1	R2	R3	R4	M1	M2	M3	M4
R1								
R2								
R3								
R4						M1		
M1								
M2								
M3								
M4		M3						R1

FIGURE 9.5 Multiplication table

o	R1	R2	R3	R4	M1	M2	M3	M4
R1	R1	R2	R3	R4	M1	M2	M3	M4
R2	R2	R3	R4	R1	M2	M3	M4	M1
R3	R3	R4	R1	R2	M3	M4	M1	M2
R4	R4	R1	R2	R3	M4	M1	M2	M3
M1	M1	M4	M3	M2	R1	R4	R3	R2
M2	M2	M1	M4	M3	R2	R1	R4	R3
M3	M3	M2	M1	M4	R3	R2	R1	R4
M4	M4	M3	M2	M1	R4	R3	R2	R1

FIGURE 9.6 A completed "multiplication table" for square symmetries

Concluding remarks

The four tasks discussed in the previous section gave a curriculum task design example of the Dual Pragmatic-Abstract Continuum. These tasks have a transitionary nature. They serve as exploratory guides for students to understand square symmetry progressively from primary school to university on the one hand, whilst on the other hand, a mathematics teacher with the abstract knowledge of group theory can design pragmatic tasks for students to learn the symmetry of the square.

Mathematical knowledge evolves continuously in a multi-perspective manner throughout a person's education. This is the main idea behind the pragmatic-abstract continuum.

The digital tool, in this case the predesigned GeoGebra files, creates a virtual transitionary boundary to foster a comprehensive experience consisting of both the pragmatic and the abstract aspects of mathematical knowledge. The GeoGebra task was designed in a way that when students are playing with the dynamic figures, they are led to go through the pragmatic-abstract continuum. The labelling of square vertices plays a significant mediating role in this aspect. The four tasks can be further refined to study different modes of thinking (e.g., geometric, algebraic, structural) and reasoning (e.g., gestural, diagrammatic, formal abstract) and their connections. The use of technology, digital or not, can be an integral part of mathematic curriculum design which resonates with the Dual Pragmatic-Abstract Continuum. Technology in education has the potential to create interactive pedagogical space for the process of knowledge exploration and construction rather than just providing students with "accurate" and pre-conceived institutionalized knowledge.

Digital technology plays the role of the boundary object in the Dual Pragmatic-Abstract Continuum in mathematics curriculum development and task design. It mediates between the concrete phenomenal world of mathematics experience and the conceptual abstract world of formal mathematics. A mathematics curriculum that is authentically real to the teachers and the students should have these two worlds integrated in complementary ways. Well-chosen digital tools/artefacts are transitory objects that possess hybrid potentials to represent both worlds; hence, with aptly designed tasks, they should inspire critical discernment for the understanding and application of the target mathematics concepts to be taught. Integrating digital technology into the mathematics curriculum and task design is about developing digital epistemic trajectories that could carry the teaching and learning of curriculum mathematics across the Dual Pragmatic-Abstract Continuum. The matching choice of digital tool/artefact and pedagogical approach is deterministic in this developmental crossing process, as this choice possesses the "mathematics carrying capacity" that makes meaningful learning achievable. In the mathematics classroom, digital technology shares a partnership with the teachers and the students, and it should not override autonomous human participation. Rather, digital technology opens up new cognitive dimensions for teachers and students to comprehend mathematical experiences.

Notes

1 Matthias Giger. Symmetrie https://www.geogebra.org/m/xt4sUqz3 Last access on 25.06.2019.
2 Symmetry of Square (Rotation) https://ggbm.at/kcqenf7k.

References

Bartolini Bussi, M. G., & Mariotti, M. A. (2008). Semiotic mediation in the mathematics classroom: Artifacts and signs after a Vygotskian perspective. In L. English (Ed.), *Handbook of International Research in Mathematics Education* (2nd ed., pp. 746–783). New York: Routledge.

Bruner, J. S. (1966). *Toward a Theory of Instruction.* Cambridge, MA: Belkapp Press.

Chan Y.C., Leung, A., & Ong, M.Y. (2017). Feedbacks and discrepancies of a physical toolkit and a digital toolkit: Opportunities and pitfalls for mediating the concept of rotational symmetry. In Allen Leung & Anna Baccaglini-Frank (Eds.). *Digital Technologies in Designing Mathematics Education Tasks – Potential and Pitfalls (Mathematics Education in the Digital Era Book Series)* (pp.259–283). Berlin: Springer.

Donevska-Todorova, A. (2017). Utilizing Technology to Facilitate the Transition from Secondary – to Tertiary Level Linear Algebra. Doctoral dissertation. Humboldt-Universität zu Berlin, Berlin. doi: 10.18452/18561 http://dx.doi.org/10.18452/18561.

Donevska-Todorova, A. (2018). Fostering students' competencies in linear algebra with digital resources. In Stewart S., Andrews-Larson C., Berman A., Zandieh M. (Eds.), *Challenges and Strategies in Teaching Linear Algebra.* Cham: Springer, https://doi.org/10.1007/978-3-319-66811-6_12.

Donevska-Todorova, A., & Weigand, H.-G. (2018). *Design principles for resources and tasks for technology-enhanced teaching and learning mathematics.* In H.-G. Weigand, A. Clark-Wilson, A. Donevska-Todorova, E. Faggiano, N. Groenbaek, & J. Trgalova (Eds.) *Proceedings of the fifth ERME Topic Conference (ETC 5) on Mathematics Education in the Digital Age (MEDA),* Copenhagen, Denmark (pp. 51–58).

Donevska-Todorova, A. (2020). Problemorientiertes Lernen zur Leitidee Raum und Form mit Nutzung von 3D Druck und AR. In Siller, H.-S., Weigel, W. & Worler, J. F. (Hrsg.). ¨ Beitrage ¨ zum Mathematikunterricht 2020. Munster: WTM-Verlag, 2020. ¨ doi: 10.37626/GA9783959871402.0 http://dx.doi.org/10.17877/DE290R-21281

Leung, A. (2017). Exploring techno-pedagogic task design in the mathematics classroom. In Allen Leung & Anna Baccaglini-Frank (Eds.). *Digital Technologies in Designing Mathematics Education Tasks – Potential and Pitfalls (Mathematics Education in the Digital Era Book Series)* (pp. 3–16). Berlin: Springer.

Leung, A., & Bolite-Frant, J. (2015). Designing mathematics tasks: The role of tools. In A. Watson and M. Ohtani (Eds.), *Task Design in Mathematics Education: The 22nd ICMI Study (New ICMI Study Series)* (pp. 191–225). New York: Springer.

Leung, A., Baccaglini-Frank, A., & Mariotti, M. A. (2013). Discernment of invariants in dynamic geometry environments. *Educational Studies in Mathematics*, 84:439–460.

Star, L. S., & Griesemer, J. R. (1989). Institutional ecology, 'translations' and boundary objects: Amateurs and professional in Berkeley's Museum of Vertebrate Zoology, 1907–30. *Social Studies of Science*, 19(3), 387–420.

Tall, D. (2004). *Thinking through three worlds of mathematics. Proceedings of the 28th Conference of the International Group for the Psychology of Mathematics Education*, Bergen, Norway, 281–288.

Vèrillon, P. (2000). Revisiting Piaget and Vygotsky: In search of a learning model for technology education. *The Journal of Technology Studies*, 26(1), 3–10.

Vèrillon, P., & Rabardel, P. (1995). Cognition and artifacts: A contribution to the study of thought in relation to instrumented activity. *European Journal of Psychology of Education*, 10(1), 77–101.

10

A FRAMEWORK FOR ANALYZING STUDENTS' LEARNING OF FUNCTION AT THE UPPER-SECONDARY LEVEL

Connected working spaces and abstraction in context

Giorgos Psycharis,

NATIONAL AND KAPODISTRIAN UNIVERSITY OF ATHENS, GREECE

Georgios-Ignatios Kafetzopoulos and

NATIONAL AND KAPODISTRIAN UNIVERSITY OF ATHENS, GREECE

Jean-Baptiste Lagrange

LDAR UNIVERSITY PARIS-DIDEROT, FRANCE

Introduction

This chapter is about a new framework for making sense of learning situations concerning functions at the upper-secondary level created by networking two theoretical approaches. The authors are researchers who share a common interest in students' approaches to functions; however, they have different research interests and theoretical backgrounds. While the third author is interested in the design and evaluation of classroom situations with broad assumptions regarding students' learning, the others pay more attention to students' evolving conceptions through their activity for the duration of the situations. Our approach to networking stems from the suggestion found in Artigue et al. (2006) that "as a research community, we need to be aware that discussion between researchers from different research communities is insufficient to achieve networking. Collaboration between teams using different theories with different underlying assumptions is called for" (p. 1242).

Two authors have collaborated previously (Lagrange & Psycharis, 2014) and implemented a "double analysis" method to analyze studies based on the use of digital environments that offer integrated geometrical and algebraic representations. This double analysis resulted in a deepened and more balanced understanding of the nature of learning situations for functions and the process of function conceptualization by students. Meanwhile, the two authors continued their theoretical work with each adopting new theoretical perspectives, *Abstraction in Context (AiC)* (Dreyfus, Hershkowitz, & Schwarz, 2015) on the one hand and *Connected Working Spaces (CWS)* (Minh & Lagrange, 2016) on the other. The work presented here goes beyond the double analysis because the first two authors collaborated with author three to design, implement and analyze a series of classroom situations as a basis for the second author's doctoral work. The whole experiment was carried out by articulating the two frameworks. At a practical level, authors one and two conducted the experiment in their school context and adopted AiC. Furthermore, they made use of Casyopée (Minh & Lagrange, 2016), a digital environment that was designed and widely experimented with in the context of author three adopting CWS. Beyond this experiment, our collaboration aimed at developing a framework for analyzing classroom situations and students' evolving conceptualization of functions. We privilege tasks involving modelling situations and the use of digital environments that offer integrated algebraic and geometrical representations of functions. Our focus is on students' transition from experiencing dependencies in non-algebraic (digital and non-digital) settings to expressing and working on these dependencies mathematically.

Approaches to functions at the upper-secondary level

The domain of functions occupies a central position in school mathematics curricula; however, reaching a suitable understanding of functions is rather difficult for many students. Research studies have evidenced students' difficulties in understanding function as covariation and linking work on magnitudes (or "quantities' measures") with mathematical functions (Thompson & Carlson, 2017). Existing research has also indicated that at both epistemological and cognitive levels, functions take on sense through experiences in many dissimilar settings. However, teaching approaches to functions at the upper-secondary level overemphasize the use of "algebraic" representations (i.e., formulas, graphs and tables) and senseless manipulation while overlooking situations, tools and resources that could facilitate students' meaningful engagement in functional thinking (Minh & Lagrange, 2016).

At the level of curriculum and textbooks, functions are often approached through tasks involving extra-mathematical situations (e.g., real-world problems) and the use of digital tools. A closer consideration, though, reveals that these tasks are mainly applications of already taught knowledge favouring the use of classical "algebraic" methods (Robert & Vandebrouck, 2014). In contrast, according to Kaiser and Schwarz (2010), authentic modelling situations are characterized by problems that are only slightly simplified and recognized by people working in a field as being problems they might meet in their daily activity. Also with regard to modelling and functions, Lagrange (2018) proposes activities in which students engage in working with models that have different relationships with reality and mathematics, paying particular attention to how to help students make connections between these models. In this chapter, we adopt a similar view that favours using modelling tasks based on authentic situations as a context for studying students' functional thinking and pays specific attention to the different settings involved in the complex path from physical context to algebra. In this path, a function exists first as a dependency between physical objects, then between geometrical objects, then between quantities and, finally, as a mathematical function.

Theoretical framework

Connected Working Spaces

Classically, activities involving mathematics and other settings are analyzed by considering that entities involved in the task appear under different semiotic representations, with each one pertaining to a field. In this multi-representational approach (e.g., Duval, 2006), activities for students in different settings are considered helpful because of the opportunities they offer for working on different semiotic representations (or *registers*) and coordinating these. In spite of the usefulness of frameworks like Duval's, for us, the "multi-representation" view is too reductively semiotic. Standing alone, this view cannot really make sense of activities that involve several fields in interaction and of their potentialities. Douady (1986) offers another framework for making sense of activities of coordinating different settings (especially mathematical ones). For Douady, a setting is constituted of objects from a branch of mathematics, a relationship between these objects and their various expressions and the mental images associated with these objects. When students solve a problem, they can consider it in different settings. Switching from one setting to another is important for the evolution of students' conceptions. For us, this framework is potentially productive in the sense that beyond representations, it puts the emphasis on mathematical contents and reasoning and their coordination across different settings. Another concern is how instruments are taken into account in students' mathematical activity. Twenty years ago, sophisticated calculators became available for students' work, and a framework was developed: the instrumental approach to the use of digital technologies to teach and learn mathematics. This approach has been inspired by research work in cognitive ergonomics, but researchers like Lagrange (1999) insisted on the intertwined development of knowledge related to the instrument and knowledge about mathematics in an instrumental genesis.

Each framework, multi-representation, coordination of mathematical settings and instrumental approach focusses on a specific dimension: the semiotic processes, the contents and reasoning and the use of instruments. We retain the idea of *Mathematical Working Spaces (MWS)* because it offers a framework associating the three dimensions. According to Kuzniak and Richard (2014), a MWS is an abstract space organized to ensure the mathematical work in an educational setting. Work in a MWS is organized around three dimensions: semiotic (symbol use, graphics, concrete objects understood as signs), instrumental (construction using artefacts, such as geometric figure, graphs) and discursive (justification and proof using a theoretical frame of reference). CWS builds on this idea of MWS by considering that in activities involving mathematics and other settings, students have to work in several working spaces and coordinate the semiotic, instrumental and discursive dimensions of these spaces. This applies especially in the approach to functions and modelling.

Abstraction in context

Building on Freudenthal's idea of "vertical mathematization", AiC defines abstraction as a process of vertical reorganization of some of the students' previous mathematical constructs within mathematics and by mathematical means in order to lead to a new construct for the learner (Dreyfus et al., 2015). In this process, previous constructs serve as building blocks for building new constructs; thus, students' construction of mathematical meanings resides in the verticality of the knowledge construction process and the depth of the resulting constructs. Sequences of problem situations can provide fruitful contexts to explore if and how new constructs operate as potential building blocks for further constructions.

AiC describes the process of abstraction by means of a model of three epistemic actions that researchers can observe and analyze: recognizing (R), building-with (B) and constructing (C). Recognizing an already known mathematical concept, process or idea occurs when a student recognizes it as inherent in a given mathematical situation and relevant to the problem situation they are dealing with. Building-with comprises the combination of existing knowledge elements (i.e., recognized constructs) to achieve a goal, such as solving a problem or justifying a solution. Constructing is carried out by integrating previous knowledge elements (constructions) by vertical mathematization to produce a new structure/construct (RBC model, Dreyfus et al., 2015). The model suggests constructing as the central epistemic action of mathematical abstraction and indicates both the process of abstraction and its outcome expressed or used by the student for the first time. Students become aware of their constructs and are more flexible in using them through consolidation, a process that is likely to occur when students build with a previously emerging construct in subsequent activities. In the process of abstraction, the epistemic actions are nested dynamically – that is, R- and B-actions are nested within C-actions, while the R-actions are nested within B-actions. Although this process describes an essentially cognitive approach, AiC attributes special attention to the role of contextual factors (e.g., resources, modelling) in learning. In studies involving authentic modelling and the use of digital tools like the present one, the analysis based on RBC-actions needs to take into account the richness of the available resources.

Combining AiC and CWS

Aiming to explore the different settings in which functions can make sense, Minh and Lagrange (2016) consider three CWS and analyze students' and teachers' evolution concerning functions in terms of how they understand the affordances provided by each of these spaces. These researchers' analysis highlights a gradual construction of personal spaces and the connections between them and implies the cognitive evolution of the subjects. However, in the CWS framework, precisely identifying subjects' cognitive evolution remains an open issue. It is then expected that identifying AiC's epistemic actions in episodes indicating the progressive construction of personal spaces will help to make sense of the subjects' cognitive evolution. Then CWS can be seen as providing a "horizontal view" of students' progress along a path from physical settings to mathematical functions, while AiC provides a "vertical reorganization view" of knowledge construction at key stages of the path. The two frameworks give some importance to "a priori analysis" in order to orient the enactment of an experimental learning situation and as a basis to make sense of the observations. Consistent with the aforementioned horizontal view, we expect from CWS an a priori analysis that precisely identifies working spaces and the three dimensions (semiotic, instrumental, discursive) in each one of these and helps to foresee students' and teacher's work and the possibility of productive connections. Consistent with the earlier vertical view, we expect from AiC an a priori analysis identifying knowledge elements, "typically concepts or strategies thought of in terms of the content domain" (Dreyfus et al., 2015, p. 191). Our research questions are as follows:

(1) How can AiC and CWS help to appreciate the potential of an implementation of an authentic modelling problem?
(2) How can combining AiC and CWS help to make sense of students' conceptualizations of function as covariation in different settings while working on an authentic modelling problem?

Method

The context

The research reported in this chapter is the first part of an ongoing classroom-based design research that aims to study meaning generation for function as covariation by 16-year-old students working in groups with concrete materials (e.g., manipulatives) and software environments to model a series of real-life situations. The experiment we present here took place in a secondary school with one class of twenty-two 11th grade students (ten groups of two or three). One researcher acted as teacher (called teacher in the chapter) and another had the role of participant observer in the classroom. The class had 14 teaching sessions (45 minutes each) over three months (one teaching session per week). At the time of the study, the students had been taught about function as correspondence (according to the curriculum), monotonicity and extreme points.

The choice of a problem as a basis for a situation

The gutter problem (Figure 10.1). People designing gutters made of a sheet of metal to channel rainwater from a roof aim to fold the sheet in order to maximize water flow. Such designers know that maximizing water flow is achieved by maximizing

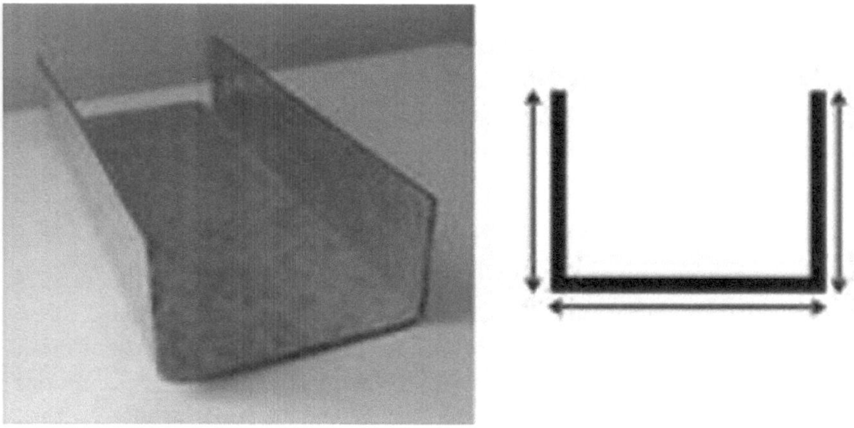

FIGURE 10.1 Gutter design

the cross-sectional area and that the sides of the rectangle (L and l) are linked by the relationship $L + 2l = C$ (C being a side of the sheet of metal). Then they use mathematics to establish that an optimal shape is for $L = 2l$ and $l = C/4$, $L = C/2$.

For secondary students, the main idea at stake is covariation emerging from authentic constraints and the conception evolves through the consideration of successive models. In considering these models, specific attention is given to the problematic nature of coupling reality and mathematics. An essential initial step of modelling concerns developing a clear understanding of the reality that is to be represented mathematically. Thus a **first model** stays close to reality: a rectangular sheet of paper can be folded in order to make a rectangular cross-section and one can appreciate sensually how the choice of the folds influences water circulation. The covariation in this model is between concrete entities (i.e., folds, flux), but the variations are difficult to appreciate because of the poor dynamicity of the model. **A geometric model** derives from the recognition that the rectangular cross-section is what has to be optimized. Dynamic geometry (DG) adds dynamicity and interactivity and helps students become aware of variations. A single free point has to be created that allows constructing the other "dependent" points of the rectangle. Constraints have to be set on this point and on the other points in order to reflect the real cross-section's constraints. This DG model involves quantities in the expression of the constraints. However, covariation is between geometric entities, points and rectangles. Thus while moving the free point, one can observe the variations of the rectangle: the rectangle is "flat" for the two limits of the variations of the free point, while it grows and decreases as the point moves from one limit to the other. The next important step is a **model involving quantification**. Thompson and Carlson (2017) consider this a rather complex passage for students. After overcoming difficulties, though, observation of quantities in this model (rectangle area, side length) should allow students a better understanding of covariation, meaning covariation between quantities rather than between geometric entities. In a **final**

model, a selected relationship between quantities (i.e., a length of the rectangle, area) can be expressed by the mathematical formalism of functions. This is again a non-obvious step, as it requires understanding the notions of independent and dependent variables and working with algebraic registers. This model allows applying classical techniques for solving the problem.

The choice of a digital environment

The Casyopée software (Figure 10.2) has been chosen because it offers both a DG window and a symbolic window with registers tabular, graphic and symbolic allowing students to work on a geometric model, as well as on the final algebraic model with the same environment. Moreover, Casyopée offers a "geometric calculations" tab: students can define independent and dependent quantities as measures in a "geometric calculations" tab and investigate their covariation. Minh and Lagrange (2016) explain the various relationships that can be made between the DG window and the symbolic window, as well as with the "geometric calculations" tab, consistent with the idea of CWS. Especially, a pair of measures in functional dependency can be exported to the symbolic window as a function that can be further treated by the students using formulas, tables and graphs. This functionality – called "automatic modelling"– is designed to help students in modelling dependencies.

The activity sequence

The activity sequence started by an introductive phase (two teaching sessions): after an introduction to the main features of Casyopée (e.g., dependencies by moving free points, definition of geometric calculations, automatic modelling), the students were presented the problem of the closest point on a parabola from a given point (Minh & Lagrange, 2016). Then in the next 12 teaching sessions, the students were engaged in modelling three realistic situations through the tasks *gutter*, *front of a store*, *oil tank*. In this chapter, we analyze the first task and its implementation (four teaching sessions).

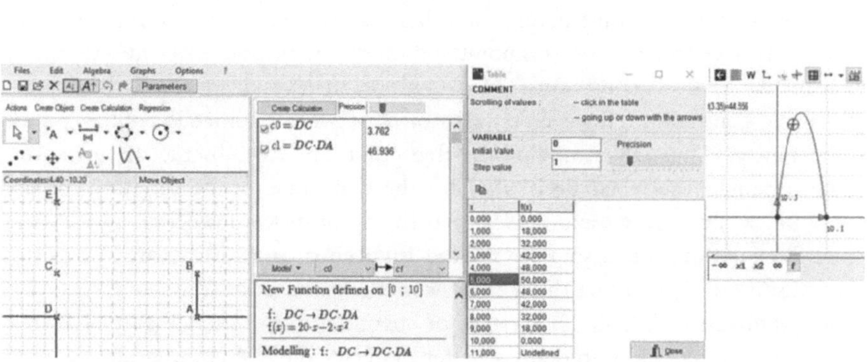

FIGURE 10.2 DG, geometric calculations, table and graph

The task presented to the students

In the *gutter* task, the students were engaged in exploring the aforementioned gutter models. The task was divided into four steps corresponding to students' activities in each model: (1) experiment with folding a piece of paper (10 cm x 20 cm) to explore and notice dependencies, (2) construct a dynamic DG figure that models the situation in Casyopée, (3) explore the use of the software to propose quantities in functional dependency and (4) obtain a mathematical function modelling the dependency and use algebraic techniques. The task can be considered a case of authentic modelling since the problem directly makes sense in a workplace field and replacing metal with paper does not oversimplify the problem. Moreover, the task is designed so that students could make sense of the situation and develop an understanding of how a range of tools (including Casyopée) can be used to provide different models.

A priori analysis of the situation

CWS analysis

We characterize four working spaces and their connections in order to foresee what students' work can be in a classroom situation based on the gutter problem.

The "sheet of paper" working space. There is a strong instrumental dimension: folding helps to approach a solution and conceptualize the covariation. The discursive dimension involves argumentation about the rectangular cross-section as representative of the flux. In the semiotic dimension, it is expected that the vocabulary will be consistent with the "real-life" situation: bottom, edge, water flow, cross-section, etc. These three dimensions of the work are linked to the realistic context, which is for us a characteristic of authentic modelling.

The "DG" working space. There is also a strong instrumental dimension in this working space created through the use of a DG environment. Building a model in this DG working space is far from obvious for students since the idea of dependency at stake in this situation underpins the construction. The semiotic dimension is in the geometrical objects (free and dependent points, segments, etc.) used to denote important elements of the gutter. As a difference with the sheet of paper, the constraints have to be expressed explicitly, using the capabilities of the DG window, and this implies adopting the associated semiotics.

Connection with the sheet of paper working space. The experience of the previous model of the sheet of paper should help students to understand and express the involved constraints. In this working space, variations are more easily observed, and this should help argumentation in the discursive dimension.

The "measure" working space. We consider the work on quantities in a specific working space due to the gap existing in students' conceptions between covariation of geometrical objects and covariation of quantities. As explained earlier, in Casyopée, quantities are handled in a "geometric calculation tab". This allows forming and numerically exploring "pre-algebraic" formulas that represent geometrical quantities. For instance, in the gutter problem, the area of the rectangle has to be expressed as the product of two lengths. Then the semiotic dimension is characterized by this "pre-algebraic" symbolism. This tab also offers the possibility of selecting two formulas (i.e., one for the independent variable and the other for the dependent variable) and checking whether the corresponding quantities are in functional dependency. Then the argumentation in the discursive dimension can go beyond covariation between geometrical objects towards dependency between quantities. The principles at stake in the argumentation deal with dependencies relative to a dynamic figure that are at the core of the idea of function: a dependency has to be expressed by way of quantities that are measurable attributes of a geometrical object, and one quantity (the independent variable) has to depend univocally on this object.

Connection with the DG working space. The measure working space introduces a new pre-algebraic symbolism and new means to work on a dependency. The action (drawing a point, observing variations, etc.), though, is still close to the work in DG. This should help students grasp how quantities with their specific symbolism are related to geometrical objects and how quantifying helps to make sense of a dependency.

The "mathematical functions" working space. This space is consistent with the "traditional" paper-and-pencil school culture involving analytic functions, algebraic notation, graphs, tables and transformations conserving equivalence. However, in this culture, the idea of function is often overlooked because of an overemphasis on algebraic manipulation. We expect that in this working space, students will reflect on representations in three registers (formulas, tables, graphs) rather than on the application of algebraic techniques. Casyopée's capabilities contribute to the instrumental dimension of the work in this space. The discursive dimension is again about the variations. However, it differs from the other spaces in that it adopts the standard mathematical way of reasoning (independent/dependent variables, maximum value), and the semiotic dimension is also the standard mathematical registers.

Connection with the measure working space. We expect that students will associate the ideas of mathematical function and of dependency between measures at the level of formalism (for instance, associating a formula like $x(l\text{-}x)$ with the product of two lengths $DC \ x \ DA$) and at the level of the understanding of variations associating the independent and dependent variables with quantities linked in a particular way.

AiC analysis

The implementation in four steps intends to make covariation appear in the four different models and the corresponding working spaces (presented earlier). In terms of AiC, the knowledge elements are related to these working spaces.

First step. Students should have to conceptualize the important elements of an authentic situation (i.e., ways of folding, volume and cross-section) by recognizing the opportunities offered by the sheet of paper working space, experimenting with the model by folding it at their own initiative and constructing an understanding of the variations (i.e., relating the gutter sides to the problem, recognizing the cross-sectional area as suitable for maximization). At each one of the next steps, the students are expected to recognize means offered in a new working space as corresponding to constructs of the previous model in order to build a new model and construct new knowledge elements relatively to covariation and functions. This recognition is expected to be feasible, thanks to the connections between the working spaces identified earlier.

Second step. In order to use the DG environment, students should have to identify geometrical objects corresponding to elements of the paper sheet model and organize these objects in order to build a DG model consistent with the paper one. The corresponding knowledge elements are (2a) the idea of independent geometrical entity necessary to build a dynamic model and (2b) the idea of covariation between geometrical entities. We shall say that the students have constructed these elements if they identify the need to use one side of the model as an independent entity, specify the corresponding restrictions for the points and express their coordinates accordingly using the available symbolic notation.

Third step. Students should first recognize the quantities at stake in the DG model (area of the rectangle, side length) and then covariation and dependency between these quantities. The knowledge elements here are (3a) covariation of quantities and (3b) formalization of this covariation as a pair of variables. We shall say that the students have constructed these elements if they work with pairs of covarying quantities in the automatic modelling tab and express their understanding of the role of independent and dependent variables and their covariation.

Fourth step. Students should conceptualize mathematical variables representing the quantities of the preceding model and a mathematical function representing the covariation of these quantities. The knowledge elements here are (4a) the idea of a function modelling a covarying pair of variables and (4b) mathematical techniques of working with this function using its standard registers (graphic, tabular, formal). We shall say that the students have constructed these elements if they create a function from a covarying pair of variables and explore the problem further using relevant mathematical registers.

Data and analysis

The collected data consists of video and audio recordings (four groups) that were fully transcribed. In the first level of the analysis, we selected episodes in which the students referred to covarying magnitudes for every different working spaces. In the second level, the selected episodes were analyzed twice according to CWS and AiC. In this chapter, we analyze the work of three groups of students (groups 1, 2, 3).

A posteriori analysis

First step: Identifying dependencies through a paper model

Observation. This episode took place during teaching session 1 after the students had experimented with a sheet of paper for some time. This is an excerpt from a discussion between two students of group 1 (S1, S2), two students of the adjacent group 2 (S3, S4) and the teacher (T).

76 S1: I think that the maximum water volume depends on the maximum volume of the metal sheet.
77 S3: [*Showing the cross-section of the gutter shown in the picture of the worksheet*] It depends on the area of this figure.
78 T: Why?
79 S3: Because the volume would be this [*showing the cross-section*], and the length of the whole thing [*gutter*]...(*does not matter*).
At this moment, S3 folds a sheet of paper with a very small vertical edge.
80 S3: What about if I fold the metal sheet like this? I want to have the maximum water volume.
81 T: The students here say that you need to maximize only the cross-sectional area.
82 S3: Actually, we need to concentrate on the changes. As the length grows, the height diminishes and the area changes. So we have to find out a proportion for which the height and length will be exactly what we need.
83 S1: Basically, the maximum product.

CWS analysis. The episode reflects students' work in the *discursive* dimension when they try to make sense of the variations at a qualitative level. In addition to referring to concrete entities (water volume, "whole thing"), students introduce geometrical vocabulary, as well as magnitudes (lengths, area, product, etc.). The three vocabularies are "blended": concrete objects, 2D and 3D geometrical elements; the lengths are not distinguished, and the magnitudes are not quantified. The students use words that make sense for them, but this use is not accurate enough to bring into play *semiotic* systems. Folding is used as an *instrument* to provide instances of the gutter adequate for the argumentation (gutter with a very small height) but insufficient to make precise sense of the variations.

AiC analysis. The episode shows students *building-with* elements of the paper model after *recognizing* the opportunities offered by the sheet of paper working space for experimenting with different dimensions and then *constructing* an understanding of the variations. When S3 folds with a very small vertical edge, she demonstrates that she *recognized* the opportunities offered by the sheet of paper to *build-with* the paper sides in order to (a) explain her idea of the relationship between the bottom and the vertical edge and their influence on the cross-sectional area and then (b) *construct* her understanding of covariation. Later on, S1 abstracts the need for a "maximum product" as a criterion for maximizing the water flow (*constructing*, line 83).

Second step: Modelling in DG

Observation. The episode is a dialogue between the teacher and students from different groups (teaching session 2). The following extract is taken from an interaction between the teacher and the group 2 students while working to construct the dynamic rectangle in the DG window (see Figure 10.2 left).

67 T: You will need one point for the lower part of the gutter and one point which describes the maximum folding. Then we need another point between these two points to describe every time the different folding, but first of all, we need to find the restriction of the construction.

68 S3: I propose to put point *D* in (0;0).

69 S4: We have to create a point *E* as (0;10) in case we fold the metal plate in the middle so that we get a segment for positioning the free point *C*.

After creating *C*, the students observed the folding in order to find an expression for the *x*-coordinate of *A*. Most of the groups attempted to find it through solving the equation x + x + y = 20 for y. Students from different groups commented, "I tried to create *A* as *(20 − 2⋆x ; 0)*, but it did not work!" "We created *A* as *(20 − 2⋆yC;0)* and it worked!" "As for us, we created *A* as *(20 − 2⋆DC;0)* and it worked also!"

CWS analysis. In order for students to take advantage of the DG possibilities, they need to identify key elements of the model as geometrical objects using geometrical notations. The teacher's intervention is crucial, insisting on the choice of points defining a rectangle, but the students also have their part: propositions for creating points *E* and *D* involving the constraints of the sheet, expression of the dependency of point *A* to point *C*. The work is in the *instrumental* dimension: adequate use of the instrument (DG) is at stake. The work also combines a *semiotic* dimension: students progressively integrate the notations of DG in order to express the *x*-coordinate of *A*. The proposition *20 − 2⋆x* by a student for this coordinate reflects an over-interpretation of the sign "*x*" as denoting "any variable".

AiC analysis. From an AiC point of view, we see here how the DG environment mediates students' identification of the correspondence between geometrical objects and elements of the paper model. Building on their experience with folding the paper sheet, the students see the need to define a rectangle through four points and to distinguish the point that "causes" the dynamic change of the construction (*recognizing*). In order to organize the objects in the DG in a way consistent with the paper, the students make faulty and successful attempts to relate the coordinates of point *A* to measures dependent on point *C* (*building-with*). In the end, different symbolizations for the coordinates of point *A* are suggested by different groups (*constructing*), indicating students' progressive coordination of their preceding sensual manipulation of the paper sheet with the available notation structures of DG.

Third step: Quantification of variations and distinction between independent-dependent variables

Observation. The selected episode refers to the experimentation of students in group 3 (S5, S6) with the geometric calculations tab of Casyopée (teaching session 3).

11 S5: Look at the area here [*pointing to the geometrical calculation DC x DA in the tab*]. We see that the maximum area is 50 and as we change this value... [*of DC*].... Okay. We cannot say that it is the maximum. If we change the point *C* in this straight line [*segment DE*], the area continuously decreases and maximizes when it [*DC*] gets its maximum value.

12 S6: Look here [*in the geometric calculations tab*], it says 50, and we have the maximum value of segment *DC*. While we move down point *C*, we see that the area is decreasing too.

After that, the students tried the automatic modelling tab, choosing the geometrical calculation *DC x DA* as an independent variable and got an error message.

80 S6: It [*Casyopée*] cannot calculate a function.

81 T: Why? What do you think about it?

82 S5: It [*the geometrical calculation*] is dependent.

83 T: So what?

84 S5: I will put here the area [*as dependent variable*] and here [*as independent one*] something that is independent of the rest of the others. That is, *DC*.

CWS analysis. In the *instrumental* dimension, the geometric calculation tab allows *continuous* observation of changes linking the movement of the free point to values of a quantity. There is a *semiotic* confusion when the students refer to the value of *DC* (DC = 5) for the area of 50 as the "maximum value of segment *DC*" since the maximum value of the length of this segment is actually 10. While students become aware of variations, they still miss the appropriate vocabulary to accurately describe them. The geometric calculation tab also provides feedback for distinguishing between independent and dependent variables. Making this distinction is not easy for students. They would prefer to consider the product *DC* x *DA* as the "main variable". In the episode, the reason why Casyopée refuses this for the independent variable is not much discussed. The students switch to another choice that they justify allusively. It would be interesting to raise this point for discussion, making the *discursive* dimension more effective. Actually, the product does not depend univocally on the position of C, and it would be useful for the students to become aware that a given value of the product is obtained for two positions of *C* (e.g., the product is 32 for both lengths 2 and 8) as a difference with the length of *DC*.

AiC analysis. By moving the point *C*, S5 *recognizes* the potentialities offered by the tab to observe continuous changes in the numerical values of the relevant measures (line 11). Then he is able to link specifically the two covarying magnitudes (line 11, *building-with*) to find a solution to the problem. Finally, S6 conceptualizes the direction of the change of these magnitudes as an abstraction (line 12, *constructing*), stating that the area decreases as the length of *DC* decreases too. As regards the students' work in the automatic modelling tab, initially, they select the area of the rectangle *ABCD* (*DC* x *DA*) as the independent variable. The provided feedback (*recognizing*, line 80) leads them to focus on the features of the selected variable and to correctly select the pair of independent-dependent variables (*building-with*, line 82). Finally, they conceptualize *DC* as a coherent element "generating" the change of the area of the rectangle *ABCD* in a rather primitive sense ("independent of the rest of the others"; *constructing*, line 84).

Fourth step: Expressing covariation through the use of variables

In teaching session 4, students of group 1 observe the changes in each column of the table of values of the function they obtained by "automatic modelling" in the symbolic window.

41 S1: From the table, we see the maximum value at 5…
42 S2: It shows the area for each value that *x* takes with the restrictions we set.
43 S1: If we change the step, it shows us the area in relation to the side *DC* that changes by 0.5. We see that 5 remains the value of the side *DC* so as to have the maximum area. We notice that for the different values of *x*, the area changes and reaches its maximum in *DC* [*equal to 5*].
44 S3 :Wait. For the various values of x, the area changes and finds a maximum for $x = 5$ with the area equal to f(5) = 50.

CWS analysis. In the *instrumental* dimension, the students coordinate observations on the table (algebraic window) and values of quantities in the geometrical calculations tab. They experiment with a smaller step (0.5) and observe behaviour of the function that is consistent with their previous observation. In the *semiotic dimension*, quantities (length of sides) coexist with algebraic symbols (x, f(…)). We observe a connection between quantities and algebra potentially productive for making sense of symbolism. While in the preceding episode students referred to the value of *DC* for the area of 50 as the "maximum value of segment *DC*", here they correctly refer to it as "the value so as to have the maximum area". It seems that working on the dependency as a mathematical function facilitates students' conception of variables.

AiC analysis. Here the students move in the direction of working with mathematical variables representing quantities that they encountered in the preceding model of the situation. They observe the variation of *DC* and its value that maximizes the area *ABCD*. By linking *DC* with column *x* of the table values (line 42), S2 helps S1 to conceptualize the variation of *DC* as the variation of the independent variable *x* (line 43, *recognizing*). Then S1 experiments with different values in the step of the table so as to determine the maximum area (line 43, *building-with*). Finally, S3 conceptualizes function as covariation by relating the changes in the two columns of the table in terms of independent and dependent variables (line 44, *constructing*). Function appears as a model for a covarying pair of variables, and students work with it through a standard tabular register.

Connections

The previous analysis reveals the progressive character of the connections students develop through building-with and constructing actions (AiC approach) in the different working spaces (CWS approach). At the second step, the students engage in the work of folding and recognize key functionalities of the DG system in order to model a variable rectangle conforming to constraints of the paper model (*"we fold in the middle..."*) and to express these constraints with the adequate notation. At the third step, the students associate a moving point in the DG figure with variations of measures (*"If we change the point C..."*) before focussing on the covariation of these measures independently of the figure. At the fourth step, the students make connections in the instrumental and semiotic dimensions between the measured space and the mathematical function space. This helps students recognize mathematical variables as representing quantities and make sense of the variations of the function as representing the covariation of measures.

This is evidence that at each step, students first connect a new working space to the former (*recognzing*) and then develop conceptualizations (*building-with* and *constructing*) inside the new working space. Beyond this step-by-step abstraction, the students in our study were able to make wider connections: for instance, a group of students commented on their solution by considering a table of the function (mathematical function working space) together with notions that exist in other working spaces (area, side etc.):

> We see in the table that the area is maximized when the coordinate of the free point C [i.e., yC] is 5. That is, we have the maximum area when one side [of the gutter] is half of the other.

Summarizing, in order to address students' conceptualizations of function as covariation in different settings, we considered a situation based on an authentic modelling task that utilizes the potential of different models, corresponding working spaces and their connections for creating meaningful abstraction processes. The transcribed episodes were subjected to a dual-lens a posteriori analyses that complement each other:

- CWS allows (a) specifying the working spaces involved in approaching functions and modelling and their connections and (b) analyzing and coordinating a triplet of dimensions of students' mathematical work in these spaces (semiotic, instrumental, discursive).
- AiC offers a fine-grained analysis of the students' constructing processes within and between these spaces.

Discussion and conclusion

By networking CWS and AiC, we introduced a framework to make sense of the implemention of an authentic modelling learning situation and students' conceptualizations of function as covariation. Next, we summarize how the two frameworks combine first in the a priori analyses and then in each one of the four steps.

In the a priori analysis, CWS provides an analytic description of the means offered to students for working on a model at each step of the situation and assumptions about connections that students can make between the models. AiC builds on this analytic description in order to identify precisely the elements of knowledge at stake and makes hypotheses about how students will learn through the connections – that is, which constructs might be observed during the process of knowledge construction. Then the combination is productive since CWS alone allows merely minimal assumptions regarding learning, but it provides a reliable structure for more precise hypotheses by AiC. In the first step, CWS insists on the instrumental dimension of the work on the paper sheet, and AiC shows how this work can be described in terms of recognizing, building-with and preparing to construct a first understanding of covariation. In the second step, CWS and AiC both use their own constructs but are consistent in noting how the students use their previous experience of the paper model to build a DG model. AiC goes deeper into knowledge construction, while CWS insists on the coordination of the instrumental and semiotic dimensions in students' work. In the third step, CWS points out difficulties for working on measures in both the semiotic and the discursive dimension, while AiC shows how the students progressed in their understanding of covariation in spite of these difficulties. In the fourth step, CWS notes that students connect quantities and algebraic entities in the instrumental and semiotic dimensions while AiC shows how this connection allows them to understand function as covariation with specific techniques and notations. As a whole, at each step, we observe processes of abstraction that at first connect a new working space to the former and then develop conceptualizations inside the new working space.

With regard to the research questions, the combination allows for the appreciation of the potential of this implementation of an authentic modelling problem in which students engage thoroughly and consider successive models from a concrete mock-up to a mathematical function. Implementation of similar tasks based on experimental situations can be informed by the insights provided by the present work in many ways. For instance, the a priori analyses may support teachers in orienting students' activities during the task enactment in the classroom by identifying different working spaces. Also, the dual-lens analysis may further sensitize task designers, teachers and researchers to (a) the diversity of models and working spaces, (b) the intended constructs in these spaces and (c) the complexity of students' constructing processes in developing the targeted conceptualizations.

As regards the contribution of each theory, the CWS allows for distinguishing the different spaces involved in the complex path from authentic context to algebra and their connections. It also allows for focussing on three dimensions (instrumental, semiotic and discursive) and their coordination in students' work. The RBC model of AiC is powerful here to make sense of students' progress, but it could not be put into operation without the structure provided by CWS. To sum up, CWS is useful in providing ("horizontally") a plurality of settings (physical context, geometry, measures, algebra) in terms of signs, instruments and modes of reasoning for students, while AiC offers concepts and expected strategies and an account of knowledge construction ("vertically") within and between these settings. This takes the analysis further and allows a deeper look at students' cognitive evolution during knowledge construction. CWS builds on the MWS theory. As we pointed out, it emphasizes the interest of offering students a variety of working spaces and the opportunity to make connections. It combines well with AiC, which helps to figure out how connections *between* working spaces contribute to conceptualization. In the MWS theory, conceptualization is considered through specific "geneses" *inside* a MWS linking epistemological and cognitive levels. In this chapter, thanks to AiC, we also observed conceptualization *inside* each of the working spaces. Thus the further step of networking could study what aspects of conceptualization are better addressed by AiC and by geneses and the possible mutual benefits.

References

Artigue, M., Bartolini-Bussi, M., Dreyfus, T., Gray, E., & Prediger, S. (2006). *Different theoretical perspectives and approaches in research in mathematics education*. In M. Bosch (Ed.), *Proceedings CERME 4* (pp. 1239–1244). Sant Feliu de Guíxols, Spain: FUNDEMI IQS – Universitat Ramon Llull and ERME.

Douady, R. (1986). Jeux de cadres et dialectique outil–objet. *Recherches en didactique des Mathématiques*, 7.2, 5–31.

Dreyfus, T., Hershkowitz, R., & Schwarz, B.B. (2015). The nested epistemic actions model for Abstraction in Context: Theory as methodological tool and methodological tool as theory. In A. Bikner-Ahsbahs, C. Knipping, & N. Presmeg (Eds.), *Approaches to Qualitative Research in Mathematics Education* (pp. 185–217). Dordrecht, The Netherlands: Springer.

Duval, R. (2006). Quelle sémiotique pour l'analyse de l'activité et des productions mathématiques? *Relime, Numero Especial,* 45–81.

Kaiser, G., & Schwarz, B. (2010). Authentic modelling problems in mathematics education. Examples and experiences. *Journal for Didactics of Mathematics*, 31(1), 51–76.

Kuzniak, A., & Richard, P. R. (2014). Spaces for mathematical work: Viewpoints and perspectives. *Relime*, 17(4.1), 17–27.

Lagrange, J.-B. (1999). Complex calculators in the classroom: Theoretical and practical reflections on teaching pre-calculus. *International Journal of Computers for Mathematical Learning*, 4(1), 51–81.

Lagrange, J.-B. (2018). *Connected working spaces: Designing and evaluating modelling based teaching situations*. In E. Bergqvist, M. Österholm, C. Granberg, & L. Sumpter (Eds.), *Proceedings of PME 42* (vol. 3, pp. 291–298). Umeå, PME.

Lagrange, J.-B., & Psycharis, G. (2014). Investigating the potential of computer environments for the teaching and learning of functions: A double analysis from two research traditions. *Technology, Knowledge and Learning*, 19(3), 255–286.

Minh, T. K., & Lagrange, J. B. (2016). Connected functional working spaces: A framework for the teaching and learning of functions at upper secondary level. *ZDM—The International Journal on Mathematics Education*, 48(6), 793–807.

Robert, A., & Vandebrouck, F. (2014). Proximités-en-acte mises en jeu en classe par les enseignants du secondaire et ZPD des élèves: analyses de séances sur des tâches complexes. *Recherches en didactique des mathématiques*, 43(2–3), 239–285.

Thompson, P. W., & Carlson, M. P. (2017). Variation, covariation, and functions: Foundational ways of thinking mathematically. In J. Cai (Ed.), *Compendium for Research in Mathematics Education* (pp. 421–456). Reston, VA: National Council of Teachers of Mathematics.

11

CHALLENGES OF MAKING SENSE OF TASKS AND AUTOMATED FEEDBACK IN DIGITAL MATHEMATICS TEXTBOOKS

Sebastian Rezat,

PADERBORN UNIVERSITY, GERMANY

Florian Schacht and

UNIVERSITY OF DUISBURG-ESSEN, GERMANY

Uta Häsel-Weide

PADERBORN UNIVERSITY, GERMANY

Introduction

Textbooks play an outstanding role in mathematics classrooms all over the world (Rezat, 2009; Fan, Zhu, & Miao, 2013, p. 636). Teachers use them for preparing their lessons (Chazan & Yerushalmy, 2014, p. 67) and for instruction in class. Students use them for learning (Rezat, 2013). Much effort is put into the development of digital textbooks, and they are increasingly present in classrooms around the world (Choppin, Carson, Borys, Cerosaletti, & Gillis, 2014). Digital textbooks have the "potential to change the textbook industry" (Pepin et al., 2015, p. 639). One of the main reasons for that is seen in the possibility to continually edit and supplement the content—and such changes "can reshape the relationships between textbook author or curriculum developer, teacher, and student" (Pepin et al., 2015, p. 639). Compared to traditional printed textbooks, the variety of digital textbooks is much broader. Pepin et al. (2015) distinguish three types of e-textbooks: (1) the *integrative e-textbook* that is traditionally authored but offers add-ons or links to learning objects that are traditionally not assumed to be part of a textbook, (2) the *evolving e-textbook* that is continuously further developed by a community of teachers or authors and (3) the *interactive e-textbook* that consists of a variety of interactive learning objects, which can be flexibly linked and combined. From this categorization, it becomes apparent that digital textbooks offer features that traditional textbooks do not comprise. These features are mostly assumed to lower costs and at the same time increase accessibility, interactivity, individualization and customization, social interactions and embedded assessment systems that can make digital materials more adaptive and assessment data more actionable (Choppin & Borys, 2017). Choppin et al. (2014) distinguish two types of digital curriculum materials based on features that contribute to these potentials: (1) individualized learning programs and (2) digitized traditional textbooks. While the potentials of the additional features of digital textbooks for the learning of mathematics are constantly emphasized, little is known about how they actually affect the learning of mathematics in practice.

In this chapter, we want to contribute to a better understanding of the effects of digital textbooks on students' learning processes. We analyze two episodes—one from a third-grade mathematics classroom and one from fifth grade—in which students work with two different digital textbooks that are currently available on the German market. Our aim is to better understand the role of the digital textbook in the learning process.

Theoretical framework

Our analysis is rooted in activity theory with its focus on the mediation of artefacts within human activity. We focus on the interactions between teacher, students and mathematics mediated by digital textbooks as modelled by the didactical tetrahedron (Rezat & Sträßer, 2012). In order to conceptualize the mediating role of artefacts, we build on both the instrumental approach (Rabardel, 2002) and the theory of semiotic mediation (Bartolini Bussi & Mariotti, 2008).

A core construct in the framework of semiotic mediation is the notion of the *semiotic potential of an artefact*. It denotes the

> double semiotic link which may occur between an artefact and the personal meanings emerging from its use to accomplish a task and at the same time the mathematical meanings evoked by its use and recognizable as mathematics by an expert.
>
> (Bartolini Bussi & Mariotti, 2008)

Thus in the framework of semiotic mediation, a distinction is made between the meanings related to solving a task and related mathematical meanings.

A comparable distinction is made by the instrumental approach. At the core of the instrumental approach is the notion of the *instrument* that the user of an artefact develops in use by both attributing functions to the artefact and applying and elaborating *utilization schemes*. Rabardel (2002) distinguishes between *usage schemes* and *instrument-mediated action schemes*. While usage schemes relate to the handling of the artefact, instrument-mediated action schemes relate to actions, which are directed at the object of the activity and mediated by the artefact.

The affordance of both frameworks is that they relate the actions carried out at the artefact to the conceptual domain. Consequently, both frameworks have been applied to technological artefacts, such as dynamic geometry systems, in which the user acts upon a semiotic representation of a mathematical object. Therefore, a close relationship is assumed between the meanings derived from solving the task and the related handling of the artefact on the one hand and mathematical meanings on the other.

In terms of digital textbooks, the situation is different. Digital textbooks often offer tasks with particular answering formats, such as multiple-choice, drag-and-drop or input of text. Therefore, the usage schemes of the artefact and related meanings derived from solving the task are not necessarily linked to a mathematical concept but challenges students with an additional set of signs to be interpreted in the course of solving the mathematical task.

Considering this particular nature of digital textbooks and building on the ideas of the framework of semiotic mediation and the instrumental approach, we distinguish two levels of signs that students have to make sense of: (1) the *artefact level*, which relates to the signs at the level of handling the artefact in order to solve the task and (2) the *content level*, which is related to the mathematical concept or meaning of the task.

A further aspect of the framework of semiotic mediation is the crucial role of social interaction for the construction of mathematical meaning. Students interact with the mathematical task and interpret mathematical signs and also interact with their peers and the teacher. They collectively interpret semiotic representations of the tasks on the artefact level and of the mathematical concept. Since "meaning and understanding in mathematical interaction cannot be directly transferred but must be seen as a personal construction" (Steinbring, 2015, p. 286), the role of social interaction, communication and discussion is crucial for learning.

Therefore, we focus on the role that the two levels of signs play in mediating the construction of mathematical meaning in social interaction amongst students. Our research question is: How are students' constructions of mathematical meaning in the context of solving tasks from digital textbooks triggered and shaped by the two levels of signs of the artefact?

Design and methods

The digital textbooks used

In our study, we use digital textbooks that are available on the German market. The reason is that we want to analyze the effect of digital textbooks on students' construction of mathematical meaning in ecologically valid settings. These textbooks neither exploit the full range of interactive tools currently available nor represent the state of the art in terms of the kind of feedback provided. However, the textbooks could be used in German schools at the moment. Therefore, our study provides insight into actual learning processes with digital textbooks that could presently occur at German schools.

In primary school, the digital textbook *Denken und Rechnen interaktiv*[1] was used. It is published by a well-established publisher of traditional textbooks in Germany.

As the first digital mathematics textbook for the primary grades on the German market, it offers interactive answering formats, such as multiple-choice, drag-and-drop or text entry. Furthermore, different types of feedback are provided. Referring to the terminology of Mory (2004), knowledge of response feedback is given after each attempt to solve the task in combination with multiple-try feedback. After two attempts to solve the task, a scaffold is accessible to the student. Following the third attempt, knowledge of correct response feedback is provided. At the end of a task sequence, summative feedback is given.

In secondary school, an unpublished version of the digital textbook *Lehrwerk Mathematik 5* (Hornisch et al., 2017) was used. The textbook has a modular structure, and it contains a variety of structural elements, such as interactive answering formats (single- and multiple-choice), animations and visualizations, drag-and-drop or text entry. After entering the solution, knowledge of response feedback or knowledge of correct response feedback (or both) is provided.

Both of the textbooks are best characterized as integrative e-textbooks according to the typology of Pepin et al. (2015).

The tasks

In the task from the primary textbook, children are asked to estimate prices of goods (a bike, in-line skates, a game console) and leisure activities (a ticket for the movie *Pirates of the Caribbean*). The given price values must be assigned to the goods by placing them in the grey fields underneath the goods via drag-and-drop (see Figure 11.1). There is no distinct matching because five prices for four goods are given. More importantly, each good seen individually could be assigned different prices. For example, a bike could cost 200 € or 400 €; even 40 € is possible if, for example, a used bike is bought at a flea market. Therefore, all prices and goods must be related to each other in order to find the expected answer.

In the task from the textbook for the secondary level, the students have to compare the sizes of different areas (e.g., a soccer field, a book or a door). This should be done by placing qualitative descriptions of sizes, such as "very big", "big", "medium-size", "small" or "very small", onto the displayed objects via drag-and-drop. As opposed to the task in the primary textbook, each descriptor of a size can be assigned multiple times. Hence the students dynamically drag-and-drop the given areas into the right order. When the task is completed, the textbook gives response feedback (on demand) about the sizes that are assigned correctly and the ones which are not (Figure 11.2).

Similar to the other task, this task requires that the students mutually relate the sizes of the different objects. However, since the qualitative descriptors of the sizes are vague and ambiguous, there is no single correct solution, but it depends on both the individual interpretations of the qualitative descriptors, such as "medium small" or "smaller than medium small", and student's individual ideas and experiences related to objects, such as "your room", "the lawn in front of your house", "area of the roof of your house".

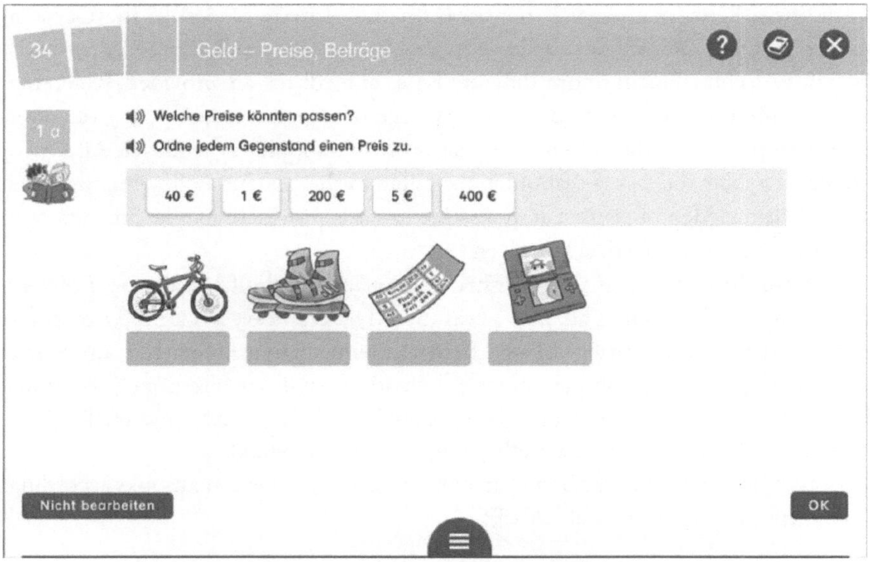

FIGURE 11.1 Task from the digital textbook (***Denken und Rechnen interaktiv*** 3, p. 34, Illustrator: Christine Kleicke © Westermann Group, Braunschweig Germany, 2011) (Translation: Heading: Money – Prices, Values; Which prices could fit? Assign a price to each object)

Estimation tasks like the ones used in the study are also present in traditional mathematics textbooks. In fact, it is likely that the digital versions are an implementation of the tasks from traditional textbooks into the digital format. Nevertheless, the semiotic potential of both representations of these tasks is different. The differences between the representation in the traditional and the digital textbook are to be found in the answering format and the automated feedback. Whereas the similar task in the traditional primary textbook allows students to relate one price to different goods, this is not possible in the digital drag-and-drop format, as implemented in the digital primary textbook. Furthermore, one of the main features of the tasks in the used digital textbooks is the immediate automated feedback that only accepts one correct solution. Due to the ambiguity of the tasks, students might develop reasonable solutions, which are evaluated as "wrong" by the digital textbooks and therefore might get the idea that there is only one correct and unambiguous solution to these kinds of tasks.

Data collection

Data was collected in a third- and fifth-grade class in two German schools. In the first episode from the primary level, the digital textbook *Denken und Rechnen interaktiv* was used in a whole-class setting in a German third-grade classroom. Each child was working with an individual tablet computer. The observation focussed on two groups of four children each. Each group was video recorded while working

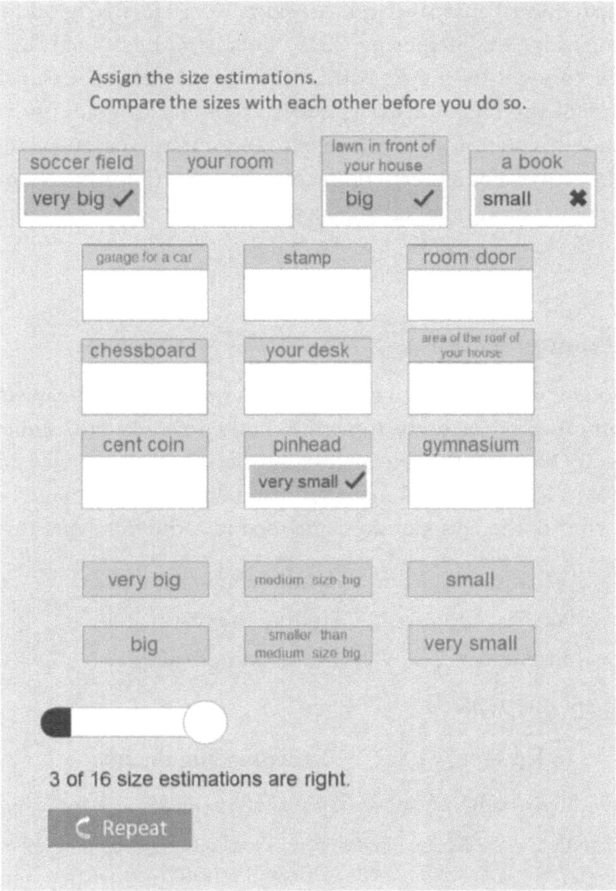

FIGURE 11.2 Screenshot and translation of a task bringing size estimations into an order

with the digital textbook for altogether six lessons. From the video recordings, one episode with a longer interaction amongst peers was chosen for the analysis in this chapter.

The data for the second episode was collected from students in grade five at a higher secondary school in Germany[2]. The data was collected in interview situations with three to four students working together with the digital textbook *Lehrwerk Mathematik 5*. The phases of group and individual work were video recorded, and the use of the digital textbook was recorded with screen-capturing software.

Methods of analysis

According to the important role of social interaction for the construction of meaning in the framework of semiotic mediation, we analyze students' interactions when solving tasks from digital textbooks. The in-depth analysis follows the rules of systematical-extensional analysis (Beck & Maier, 1994), focussing on the epistemological learning processes (Steinbring, 2005). Following Beck and Maier (1994), we choose crucial episodes for a systematical-extensional analysis. The following two episodes represent the range of the episodes according to grade (primary school, secondary school), situation of use (solving tasks, interpretation of the feedback), digital medium (tablet, computer), cooperative learning (informal cooperation situation, formal cooperation moderated by a teacher).

Analysis

Episode 1: Primary school

The first episode is video recorded in a classroom situation in primary school (grade 3). After a short introduction to the topic, the children are asked to estimate the prices of goods (Figure 11.1). Although each child uses his or her own tablet with the digital textbook, a lot of informal interaction and collaboration could be observed.

The transcript[3] of the episode can be divided into the following three parts.

1–11	"What should I do?" – Understanding the task
12–21	"You have to press". – Finding the first match
22–33	"I think five Euros". – Estimation of values

Analysis of episode 1, part 1:

1–11	"What should I do?" – Understanding the task

The episode begins when Max starts to solve the task.

1	Max	*(Taps a number of times with his finger on the grey input screen on the right.)* But what should I do now? What should I do here *(points at his tablet computer)*?
2	Sarah	You have to guess; you have to guess how much.
3	Max	Here *(points at his tablet)*.
4	Sarah	You put the two hundred here, for example *(points at the 200 € and at the grey input screen on the right-hand side)*.
5	Max	Ok, ok, I wanted to do so.
6	Sarah	But a ticket, a ticket for cinema or something like that *(points at the ticket)* costs five Euros, for example.
7	Max	Mmm.
8	Sarah	Or one Euro.

The interaction between Max and Sarah starts with a question. Max asks Sarah what he has to do. It is not absolutely clear if he asks for help for the technical handling of the textbook or for a content-related explanation of the task. It can be supposed that he asks for help with the technical handling because of his tapping on the input screen before asking. His remarks and gestures in T. 3 and T. 5 might also be interpreted as hints for technical questions. Perhaps he does not know that he has to drag the prices and drop them onto the grey fields but wants to enter the price via the keyboard.

Sarah answers content related, and she explains that the values have to be estimated. She uses the expression "guess" and does not give any explanation about how to estimate the values. Because of the position of the prices and the figures, it is possible that she matches prices and figures based on their position in the task (price on the right to figure on the right). However, her examples are also reasonable from a mathematical point of view.

In both turns, Sarah uses the expression "for example". This could be a hint that she does not focus on the particular prices but rather explains more generally that objects and prices need to be assigned. However, she points out that she does not feel certain about the costs for the cinema ticket. Considering that, she seems to estimate the value herself. The uncertainty in terms of the costs of the ticket could be caused by the picture in the task because the ticket is not easily identifiable as a cinema ticket ("ticket for cinema or something like that"). This could be a hint that she argues content related.

Even though Sarah speaks of guessing and does not explain her references for the values, she seems to have a concept of money values. If Max's question was content related, her explanation could help Max solve the task.

Analysis of episode 1, part 2:

12–21		"You have to press". – Finding the first match
12	Sarah	This Internet, for example (*points at the game's console*). My father bought it; this costs two hundred Euros for sure.
13	Max	How much?
14	Sarah	No, or four hundred.
15	Max	No two hundred. I was given, I was given this once [as a present].
16	Sarah	Okay, then you have to press two hundred certainly (*points with a pen towards the prices*).
17	Max	(Presses 200 € two times) Huh?
18	Sarah	Then you have certainly to (*tries to drag the price 200 € on Max's tablet to the grey input screen below the console figure*) Huh (*presses repeatedly on the prices*)?
		(*Points at the task formulation*) Which price could fit? Match objects and prices (*drags 200 € at the input screen below the console figure; the 200 € disappear from the price line*).
19	Max	Ah. Okay.
20	Sarah	I'm clever, aren't I?
21	Max	Yes, you are.

Sarah turns back to the first example and gives a reason for her estimation of the price of the console now. She mentions that her father has bought a console for 200 €. It is not clear if she remembers the actual costs of a real purchase or if she wants to express that her father has a lot of money. Max seems to agree with her because he presses the 200 € button.

Max's action shows again that he does not know how to enter the solution. Sarah seems to suggest that he should drag, at first verbally and then by dragging herself. However, at first glance, she does not succeed either. After re-reading the task, she tries again, and this time succeeds.

The first two parts of the episode already illustrate some challenges on the *artefact level* of tasks in the digital textbook. The children have to develop usage schemes related to the given task. They have to recognize how to enter the solution and must carry out the drag-and-drop procedure correctly.

On the *content level*, the (figural) sign must be interpreted (cinema ticket, game console). Even if the figures are not abstract mathematical signs but illustrations, it becomes clear in the discussion that the children have to interpret the represented objects. Finally, the mathematical content of the task must be comprehended (i.e., the value of the objects must be estimated correctly).

Analysis of episode 1, part 3:

22–33 "I think five Euros". – Estimation of values

In the following part, Max starts to assign the next value on his own. At first, he seems to be confused because the appearance of the task has changed, and "200 €" cannot be found amongst the prices anymore. Then he finds the price he is looking for and drags it towards the grey field. At this moment, he seems to start thinking about the best matching goods, and his finger alternates between the in-line skates and the cinema ticket.

22	Max	Well, a table tennis table (..) Where have the five Euros gone? Ah, here they are (*puts his finger at 5 €*). Okay, I think five Euros (drags the price 5 € towards the "cinema ticket" and the "in-line skates". (*Alternates with his finger between these two figures and stops below the in-line skates*).
23	Sarah	In-line skates cost a lot.
24	Max	I know.
25	Sarah	A ticket costs [or] a bike costs certainly five Euros.
26	Max	Okay.
27	Sarah	My, my dad buys bikes for five Euros or so, too.
28	Max	Okay (*drags the price of 5 € back to the line with the prices and drops it there*).

29	Sarah	Because some people are poor.
30	Leonie	I already know the answer (*stands up and bows her head over Max's tablet*).
31	Max	That is a little to.
32	Leonie	This is forty (*points at the in-line skates*), this is four hundred (*points at the bike*) and this is (*points at the game console*) and this is five Euros (*points at the cinema ticket*).
33	Max	Mmm aha, thanks (*drags and drops the prices below the figures*).

Before Max decides, Sarah verbalizes that in-line skates cost a lot. She seems to recognize Max's uncertainty and wants to help him. Probably, she wants him to follow her suggestion from the beginning. Anyway, Sarah seems to have a concept of price values. This assumption is disturbed by her statement that a bike costs five Euros. In this part, the confusion is not resolved because the content-related discussion between Max and Sarah is interrupted by Leonie. Max does not seem to agree with Sarah, but he also does not make a different suggestion. He drags the questionable price back to the price line. Then Max assigns the prices to the goods as suggested by Leonie. This solution is evaluated as correct by the digital textbook, and they proceed with the next task.

In this part of the episode, the *content-related* discussion dominates. Max and Sarah negotiate which of the goods probably costs 5 €. Max's considerations are visible in his actions, while Sarah argues verbally. Altogether, three objects are seriously suggested that cost 5 €. The episode shows the difficulties of children in estimating price values (Franke & Ruwisch, 2010).

On the *artefact level*, the interactive behaviour of the digital textbook accompanied by the variation of the task appearance challenges Max. It is not clear if his irritation is caused by the unexpected change of the appearance of the task or by a personal uncertainty.

Summary of episode 1

On the *content level* of the interaction, students estimate the money values of the given goods. They consider different possibilities. Sarah's reasoning is based on her everyday experiences. The children seem to focus on one good at a time and try to find the correct price value for it. They do not seem to consider the relation between *all* given goods and *all* given price values. However, this would be necessary in order to find the expected solution.

On the *artefact level*, two aspects were identified. First, the children try to find the correct usage schemes for the task. The children discuss which scheme should be used and how it is to be performed. Second, an irritation caused by the interactive behaviour of the textbook is verbalized.

Second episode: Secondary school

The following episode shows three students working together on the task shown in Figure 11.2. The students have worked on the task presented earlier and ordered the objects according to their sizes by assigning qualitative descriptors of sizes by drag-and-drop. The two male students, Oliver and Henry, worked together on the task, the female student, Emily, worked on her own. The episode shows a discussion that occured after the students had received the automated feedback from the digital textbook.

The transcript[4] of the second episode can be divided into the following three parts.

1–11	"I was better with my logic". – Interpreting the solutions of the peers
12–17	"They meant a different cent coin". – Making sense of the feedback
18–24	"The computer was wrong". – The (hypothetical) logic behind the feedback

The feedback provided by the digital textbook shows that Oliver and Henry have 2 out of 13 answers correct in their joint solution (marked as green on their screen), and Emily has six correct answers (marked as green on her screen). The interviewer initiates a discussion amongst the students by asking what the students recognize. Following that question, the students try to interpret the feedback provided by the textbook and their solutions.

Analysis of episode 2, part 1:

1–11 "I was better with my logic". – Interpreting the solution

The following figures (Figure 11.3a and b) show the solutions of the students after the feedback was given.

1	I	What do you recognize now?
2	Henry	That is uhm.... Now, what is the size of the room? Small?
3	Emily	See!?
4	Oliver	The room is big.
5	Emily	Maybe you are not that bad.
6	Oliver	The room is big.
7	Henry	Oh my god. Two out of 13.
8	I	So.... What have we seen now?
9	Emily	That I was better with my logic.
10	Henry	That the computer was wrong.
11		*Laughing*

The episode starts with a question by the interviewer who wants to know what the students have recognized. First, Henry asks Emily about the size she had assigned

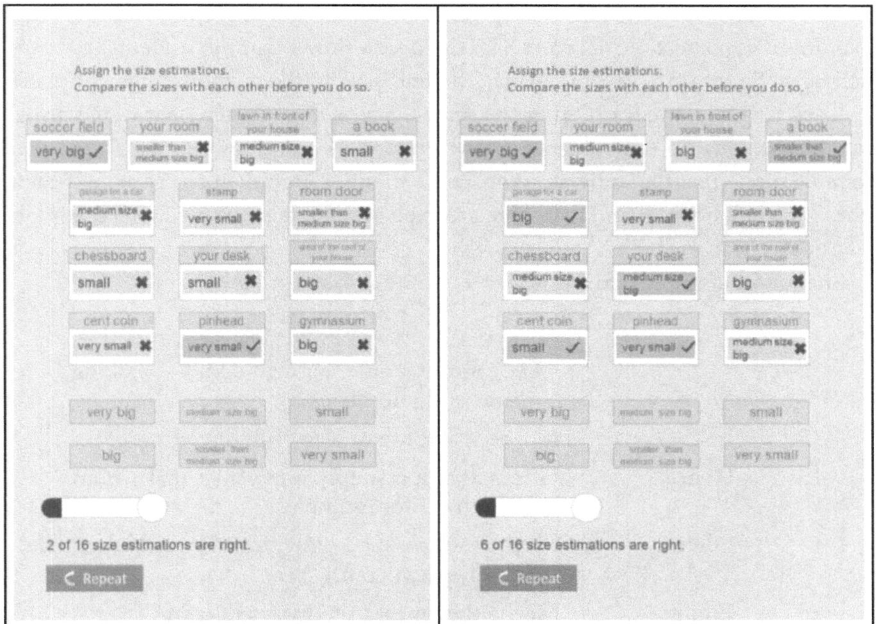

Oliver's & Henry's solution after the feedback was given.	Emily's solution after the feedback was given.

FIGURE 11.3 The solutions of the students Oliver and Henry, as well as Emily, after feedback was given.

to the room. In both cases, the students' solutions were wrong (Oliver and Henry: smaller than medium size, Emily: medium size). In T. 2, Henry asks what the actual size of the room would probably be. Oliver supposes that it shall be big (T. 4 and 6). In that situation, the interviewer asks again what the students recognize. In T. 9, Emily answers that she was better with her logic, and Henry replies that the computer was wrong (T. 10).

The episode documents a process of the students' work in the light of the given feedback (see example 1 (T. 9) and example 2 (T. 10)). In both cases, the students justify their answers in the light of the given feedback. Emily explains the fact that she had six answers correct, whereas Oliver and Henry only had two answers correct. She explains this result by claiming that she was better with her logic (T9). Oliver and Henry explain the same result differently by claiming that "the computer was wrong" (T. 10).

On the *artefact level*, students immediately understand the feedback of the digital textbook. There is no discussion about the interpretation of the signs denoting correct or incorrect solutions. The discussion relates to the interpretation of the meaning of the feedback on the content level.

On the *content level*, the students estimate the size of the room by referring to the choice of sizes. One of the challenges of the task format is that they have to make sense of their solution in light of the given feedback, which only shows if the answer is approved as correct or incorrect. The discussion shows that the students interpret such signs (Steinbring, 2005) in many different ways. In the previous interaction, the feedback is quite complex to interpret since many of the assigned sizes were wrong, and the students do not easily conclude what the correct answer would be. The example shows that the students compare the mode of argumentation amongst each other ("logic is better") and that there are explicit references to the digital textbook ("computer was wrong").

Analysis of episode 2, part 2:

12–17		"They meant a different cent coin". – Making sense of the feedback
12	I	Maybe.... How can we verify that?
13	Henry	Uhm...that it is simply impossible, that...uhm... that, that...for example.
14	Oliver	Maybe they meant a different cent coin other than the one-cent coin.
15	Emily	Maybe they meant the five-cent coin.
16	Oliver/Henry	Fifty.
17	Emily	Yes, fifty.

In the second part of the episode, the students' interaction focusses on the interpretation of the given feedback in the light of their own solutions. The interviewer continues and asks how the students could verify the assumption that Emily's logic was better and that the computer was wrong (T. 12). Henry states that the result is "simply impossible" (T. 13). In reaction to Henry, Oliver suggests that one possible interpretation of the mismatch between their own solution and the computer solution is that "they meant a different cent coin" (T. 14). By saying "they", the students refer to the hypothetical textbook authors of the task. These persons could have had a solution in mind that differs from the solution that the students came up with. Following this idea, the students suggest different coins that could be plausible (T. 15–17).

On the *content level*, the students refer to the size of the coin and the room. This explanation challenges the differences between the individual solution and the computer solution by referring to the real context and related multiple meanings of the word "cent coin" (a different cent coin, a 5-cent coin, a 50-cent coin, cf. T. 14–17). Hence, the students explain that if the hypothetical authors would have meant a 50-cent coin, the computer solution would be plausible.

On the *artefact level*, differentiated and multiple meanings are ascribed to the signs in the textbook. The feedback triggers students' thinking about the meaning of the words used in the textbook.

Analysis of episode 2, part 3:

18–24		"The computer was wrong". – The (hypothetical) logic behind the feedback
18	Henry	Yes, but, but still the stamp…right…
19	Emily	Look…
20	Oliver	…is bigger than the cent coin.
21	Emily	There you see your solution. When you look at your solution, then the only possibility is that you were wrong.
22	Henry	*(Points at Emily's monitor)* Yes, but…. [a] cent coin is small. Then the stamp should be bigger. Then this might actually be — if it is not very small — either small because it is bigger. So, then it would be smaller than medium size. But the book is already that size. That doesn't fit at all.
23	I	Right.
24	Henry	So, the computer is wrong.

In the third part of the episode, the students interpret the given feedback from a different perspective. In T. 18–20, they refer to the real objects and compare the sizes of a stamp and a cent coin. Although this seems to be plausible to Emily, she says to Oliver and Henry that the only possibility is that they were wrong (T21). Henry is challenged by that utterance and then tries to reconstruct the logic behind it (T. 22). The reconstruction of Henry's argumentation is as follows (see T. 22):

The cent coin is small. → The stamp is bigger. → The stamp has to be smaller than medium size. → The digital textbook is wrong because the book is already smaller than medium-size. (T. 22)

Hence, he accepts the authority of the given solution and tries to understand it by reconstructing the logic behind it — supposing that it is logically structured (T. 22). In this scene, the student reconstructs a hypothetical logic behind the given feedback.

On the *content level*, the scene reveals a form of meta-argumentation in which the student analyzes the computer solution with respect to the content-specific underlying logic. This logical structure itself is then compared with the logic that led to their commonly determined solution. Reasonable arguments are presented for the correctness of the students' solutions.

On the *artefact level*, the signs of the feedback still serve as a reference for the justification of the students' solutions.

Summary of episode 2

The episode exemplifies how the feedback by the digital textbook affects the social construction of meaning.

On the *content level*, the interaction focusses on the interpretation of the students' solutions in relation to the given feedback. Hence, the students interpret their own solutions and the given feedback, as well as the relationship between both. Modes of interpretation within the interaction are that the students use references to real contexts or assume a hypothetical textbook author (part 1), they analyze a (hypothetical) logical structure behind the "correct" solution of the task (which they do not know then) and compare it to the logical structure of their own solution (part 2) or they judge the mode of argumentation itself without any relationship to the context of the task (part 3).

On the *artefact level*, the signs of the feedback provoke and trigger the discussion on the content level because students are trying to make sense of the feedback they received.

In both cases, students' discussions were not taken up by the teachers afterwards.

Discussion and conclusion

In both episodes, students are working on tasks that require estimation. As we have shown in the analysis of the tasks, neither task has unambiguous answers. Nevertheless, both digital textbooks only accept one correct solution and accordingly provide knowledge of response feedback. Even if a different solution may be plausible, the textbook will give negative feedback. While in the episode from the third grade, students try to find the correct solution, the students in the episode from the fifth grade try to make sense of the (negative) feedback they received from the digital textbook. Our analysis of the two episodes revealed how in both cases the discussions of the students are triggered by students' interpretations of signs on the artefact level. In the episode from the third grade, the starting point of the discussion is Max's question about what he has to do. Our analysis shows that it is not clear whether this question is related to the content level or to the artefact level, which in this case is related to the handling of the answering format (drag-and-drop). In the episode from the fifth grade, the interviewer initiates a discussion in which students try to make sense of the feedback. The students provide justifications, which could explain why their solution deviates from the expected solution of the textbook without accepting that their own solution is wrong. This is indeed an interesting observation regarding the authority that is widely ascribed to textbooks (Luke, de Castell, & Luke 1989). In this case, the students seem to not accept the authority of the textbook but try to justify their deviation from the expected solution. This might indicate that less authority is ascribed to digital textbooks compared to traditional ones.

The analysis shows that in both episodes, the interpretation of signs on the artefact level (answering format, feedback) imposes additional challenges on the students. On the one hand, students try to understand how to handle the textbook, and on the other hand, they try to make sense of the feedback they get. At first glance, students' attempts at making sense of signs on the artefact level seem to provoke and trigger discussions on the content level. In the first episode, students start discussing the price values of the goods, and in the second episode, they start comparing the sizes of the objects again. This could be interpreted as an affordance of digital textbooks. However, although the students present interesting arguments in their discussion, these discussions do not seem to lead to an advanced understanding of the mathematical content. In the first episode, the discussion ends when another student, who already solved the task correctly, interrupts and simply prompts the correct solution. In the second episode, the students conclude that the feedback of the textbook must be wrong, and the conflict between their solution and the expected solution by the textbook is not resolved. Therefore, in the latter case, the signs on the artefact level seem to have a stimulating affective nature, and the students seem to feel the need to justify their solution against the provided feedback of the digital textbook. Furthermore, students ascribe a certain rationality to the feedback they get from the textbook and try to reconstruct this rationality.

Our analysis shows that the design of tasks and related feedback in digital textbooks requires careful consideration. The goal of the estimation tasks used in our study is that students mutually relate different objects according to their magnitudes. This requires a justification of the mutual relations, which are negotiable. Stimulated by signs on the artefact level, students present multiple arguments and more differentiated interpretations of the signs of the artefact, but since students do not get feedback during their discussions, they cannot evaluate their arguments and interpretations.

Consequently, these discussions do not lead to a shared mathematical meaning but to the conclusion that the computer must be wrong. Hence the use of such tasks with feedback needs careful handling in classroom situations. In both cases, it would be the task of the teacher to lead students' discussions to a satisfactory outcome by questioning and evaluating the presented arguments. Therefore, the digital textbooks with automated feedback in our cases do not take any responsibility from the teacher. Just as with tasks from traditional textbooks, teachers need to be aware of the goals of tasks they assign to students and how to achieve these goals. Therefore, it is questionable if digital textbooks and their affordances even change the role of the teacher within the didactical tetrahedron. Additionally, teachers need to be aware of the possible effects of automated feedback on students. In tasks where, on the one hand, multiple solutions might be plausible, but on the other hand the automated feedback only accepts one correct solution, students' attempts to make sense of the mathematics need to be resonated by the negotiation of meaning in the classroom initiated by the teacher. This should result in a shared mathematical meaning.

Notes

1 https://grundschule-interaktiv.westermann.de
2 In most states in Germany, secondary school starts at grade 5.
3 The original transcript is in German and was translated by the authors.
4 The original transcript is in German and was translated by the authors.

References

Bartolini Bussi, M. G., & Mariotti, M. A. (2008). Semiotic mediation in the mathematics classroom: Artifacts and signs after a Vygotskian perspective. In L. D. English (Ed.), *Handbook of International Research in Mathematics Education* (pp. 750–787). Mahwah: LEA.

Beck, C., & Maier, H. (1994). Zu Methoden der Textinterpretation in der empirischen mathematikdidaktischen Forschung. In H. Maier & J. Voigt (Eds.), *Verstehen und Verständigung. Arbeiten zur interpretativen Unterrichtsforschung* (pp. 43–76). Köln: Aulis.

Chazan, D., & Yerushalmy, M. (2014). The future of mathematics textbooks: Ramifications of technological change. In M. Stocchetti (Ed.), *Media and Education in the Digital Age: Concepts, Assessments, Subversions* (pp. 63–76). Frankfurt: Peter Lang.

Choppin, J., & Borys, Z. (2017). Trends in the design, development, and use of digital curriculum materials. *ZDM Mathematics Education*, 49(5), 663–674. doi:10.1007/s11858-017-0860-x.

Choppin, J., Carson, C., Borys, Z., Cerosaletti, C., & Gillis, R. (2014). A typology for analyzing digital curricula in mathematics education. *International Journal of Education in Mathematics, Science and Technology*, 2(1), 11–25.

Fan, L., Zhu, Y., & Miao, Z. (2013). Textbook research in mathematics education: Development status and directions. *ZDM Mathematics Education*, 45(5), 633–646.

Franke, M., & Ruwisch, S. (2010). *Didaktik des Sachrechnens in der Grundschule*. Heidelberg: Spektrum.

Hornisch, B., Kehret, D., Kern, F., Moser, M., Ottersbach, J., Rink, R., Roder, U., Schreiber, I. M., Tews, W. (2017). *Brockhaus Lehrwerke: Mathematik 5. Klasse. Größenvergleich von Flächen*. Unpublished textbook. München: Brockhaus.

Luke, C., de Castell, S., & Luke, A. (1989). Beyond criticism: The authority of the school textbook. In S. de Castell, A. Luke, & C. Luke (Eds.), *Language, Authority and Criticism* (pp. 245–260). London: The Falmer Press.

Mory, E. H. (2004). Feedback research revisited. In D. H. Jonassen (Ed.), *Handbook of Research on Educational Communications and Technology* (2nd ed., pp. 745–783). Mahwah: Lawrence Erlbaum Associates.

Pepin, B., Gueudet, G., Yerushalmy, M., Trouche, L., & Chazan, D. (2015). E-textbooks in/for teaching and learning mathematics: A potentially transformative educational technology. In L. D. English & D. Kirshner (Eds.), *Handbook of International Research in Mathematics Education* (3rd ed.). New York: Routledge.

Rabardel, P. (2002). People and technology: A cognitive approach to contemporary instruments. Retrieved from https://halshs.archives-ouvertes.fr/file/index/docid/1020705/filename/people_and_technology.pdf.

Rezat, S. (2009). *Das Mathematikbuch als Instrument des Schülers: Eine Studie zur Schulbuchnutzung in den Sekundarstufen*. Wiesbaden: Vieweg+Teubner.

Rezat, S. (2013). The textbook-in-use: Students' utilization schemes of mathematics textbooks related to self-regulated practicing. *ZDM Mathematics Education*, 45(5), 659–670. doi:10.1007/s11858-013-0529-z.

Rezat, S., & Sträßer, R. (2012). From the didactical triangle to the socio-didactical tetrahedron: Artifacts as fundamental constituents of the didactical situation. *ZDM Mathematics Education*, 44(5), 641–651. doi:10.1007/s11858-012-0448-4.

Steinbring, H. (2005). *The Construction of New Mathematical Knowledge in Classroom Interaction – An Epistemological Perspective*. Berlin: Springer.

Steinbring, H. (2015). Mathematical interaction shaped by communication, epistemological constraints and enactivism. *ZDM Mathematics Education*, 47(2), 281–293.

12

TECHNOLOGY INTEGRATION FOR MATHEMATICS EDUCATION IN DEVELOPING COUNTRIES, WITH A FOCUS ON INDIA AND MEXICO

Ana Isabel Sacristán,

CENTRE FOR RESEARCH AND ADVANCED STUDIES (CINVESTAV), MEXICO

Jeenath Rahaman,

TATA INSTITUTE OF FUNDAMENTAL RESEARCH, INDIA

Suchismita Srinivas and

TATA INSTITUTE OF SOCIAL SCIENCE, INDIA

Teresa Rojano

CENTER FOR RESEARCH AND ADVANCED STUDIES (CINVESTAV), MEXICO

Background and introduction

Most of the research literature on technology integration in mathematics education comes from developed countries; there is, however, a digital and educational divide insofar that a large part of the world's population doesn't have the material and human resources of developed countries. Julie et al. (2010), in the 17th International Commission on Mathematical Instruction (ICMI) Study book, did address issues of access and implementations of technology for mathematics education in different regions of the world, most involving developing countries, such as African and Latin-American countries, and concluded,

> Conditions of schooling, social stratification, even within early developing countries, and political conditions mitigate strongly against all students benefiting from the possibilities for meaningful learning of mathematics which digital technologies have to offer. Despite the possibilities offered by technology [...] such as increasing the educational options for the marginalized sectors of society, many of these are not yet widely seen, and there is still a need to reformulate the educational approaches when using new technologies. To realise this

> ideal will require thoughtful navigation, action and support, in the first instance, from the global mathematics education community.
>
> (Julie et al., 2010; p. 381)

Ten years later, a need to revisit the current situation in developing countries is called for. Grant and Eynon (2017), in the general context of technology-enhanced learning, have pointed out how the issue of the digital is not going away, with possibly wider inequalities in access, use and experiences. In this chapter, we seek to provide examples of experiences from our own countries (India and Mexico) in order to give insights into the issues faced in developing countries and how these are coped within the particular context of mathematics education.

Technology integration in India and Mexico

India is the second-largest country in the world (behind China), and Mexico is ranked 11th, with populations in 2019 of over 1.35 billion[1] and over 125 million[2] people, respectively, thus provide a large representation of the world population. Both of these countries are very diverse in social and economic terms, with also big geographic disparities (e.g., with rural areas lacking opportunities and infrastructure). According to a recent report by the Organisation for Economic Co-operation and Development (OECD, 2017), approximately 65% of young adults in India (and China) and 50% in Mexico (along with Costa Rica, Indonesia and South Africa) did not finish high school, and 46% of adults in India and 14% in Mexico did not finish primary school. These indicators give a panorama of the general educational situations in both these countries and thus contextualize the difficulties in considering technological integration in under- and developing countries. We will now discuss the issues of technology integration in general and for mathematics education for the cases of India and Mexico. We focus on state or government schools because, in both countries, there is a big divide between government and private schools.

A. The case of India

India has over 250 million schoolchildren studying in more than 1.5 million schools (MHRD, 2019). There is a huge diversity in the type of schools, ranging from elite private schools at one end, to low-fee private and government schools at the other (EQFI, 2015). The largest proportion – close to 70% – are government schools[3] (ESAG, 2018) catering to the socio-economically less advantaged sectors.

A.1. Main challenges in mathematics teaching and learning in India

A major issue facing school education in India is the huge socio-economic divide in access to quality teaching and resources (Velaskar, 2010). Improving access to mathematics for students who are traditionally disadvantaged is foregrounded in our discussion on the priority and scope of technology integration in India.

Some specific challenges in government schools are extremely high pupil-to-teacher ratio (PTR) – making it difficult for one teacher to orchestrate large classes – low student motivation and teachers' high workloads. In terms of resources, these schools rely on textbooks and chalkboards (with little or no access to digital technologies), while the medium-to-high-fee private schools have the means and access to multiple resources, including digital ones.

In terms of pedagogical practices, traditional Indian classrooms are highly teacher-centric, with teachers having a transmissive view of teaching (Kumar and Subramaniam, 2013) and promoting rote learning. Thus students seldom get involved actively in problem solving, in sharing original solutions, in building mathematical understanding (at odds with the intended curriculum) or in learning from errors. In fact, teachers are intolerant of errors in mathematics classrooms, with the common response being to show the correct procedure (Kumar & Subramaniam, 2013). In general, mistakes invite censure, sometimes even corporal punishment, which Portela and Pells (2015) found to be negatively correlated with math scores and students' agency. These practices lead to low student motivation, student learning levels below par for their grade level and very high dropout rates (in some cases over 50%), particularly in the first year of high school. Some educators (e.g., Khan, 2015) have blamed fear and failure in mathematics as one of the issues leading to such high dropouts in secondary schools.[4]

Another issue is a singular focus on (paper-and-pencil) examination "performance", which promotes rote learning, limits any scope of authentic exploration or engagement in mathematics and further marginalizes socio-economically disadvantaged students (D'Souza, 2019).

We now examine technology integration in India and discuss how technology-integrated initiatives are trying to specifically address some of these challenges.

A.2. A panorama of technology integration in primary and secondary schools in India

While many developed nations have focussed on using technologies with a distinct curricular and pedagogic vision aimed at fundamentally changing mathematics education in the classroom (Trouche et al., 2013), in contrast, policies related to the use of information and communication technologies (ICT) in Indian schools have always been somewhat casual: stressing the use of technology per se rather than *how* to use it for better learning. In the absence of a larger vision, technology use in India thus far has catered to existing curricula and pedagogic practices rather than challenging them (Tata Trust Report, 2018; Kasinathan, 2009).

The national policy discourse in India on using technology in education started in 1986, aiming at distributing computers in some (senior secondary government) schools, which happened between 1993–1998. In 1998, accessibility to low-cost Internet and computers was recommended. In 2000 and 2001, TV and radio channels were devoted to education, with an education satellite (Edusat) introduced in 2004 to reach the diverse socio-economic and geographical areas of India. The national policies that came into effect in 2000 and 2005 endorsed computer-based learning in the curriculum and the need for teacher professional development (NCERT, 2000; NCERT, 2005). Later in 2010, new recommendations were made, including the provision of a separate ICT teacher in schools (NITI Ayog, 2015). In 2013, the National Repository of Open Educational Resources[5] was launched, hosting resources in different multimedia formats in many subjects for all grades. More recently, the draft of the National Education Policy (NEP, 2019) has extolled the use of technology for education, emphasizing the use of free software to increase accessibility but recognizing the need for long-term teacher professional development.

As can be seen, the Indian government initiatives changed from recommending inclusion of technology in education, to seeking making technology reachable to a wider audience. The focus has mostly been on hardware and infrastructure – i.e., setting up computer labs in government schools. In this respect, government recommendations (NPICTSE, 2012) stipulate that – to begin with – each school should be equipped with at least one computer laboratory with at least ten networked computers accommodating 40 students at a time (see Figure 12.1). The computers have pre-loaded generic content, videos, etc.

In practice, however, ground realities are quite different from those recommended by the government schemes and policies: there are a number of challenges ranging from hardware and infrastructure issues to lack of Internet access, to availability of good open-source resources in local languages. Governments are not well informed about what to buy, often acquiring outdated hardware/software, or they are seduced by companies that push for state-of-the-art resources that are unaffordable for large-scale implementations. Thus hardware is limited and seldom updated (particularly in rural schools), which often means that the software used is the one that is available and runs on that existing hardware; also, the number of students in classes is very large, often over 50, and students cannot be accommodated in a small-size computer lab.

Even in schools where infrastructure challenges are relatively mitigated, actual use of technology is limited due to other reasons. In particular, since teachers do not receive much support, in many schools, the computers have remained locked up or at least unused for educational purposes (some have even been stolen); in others,

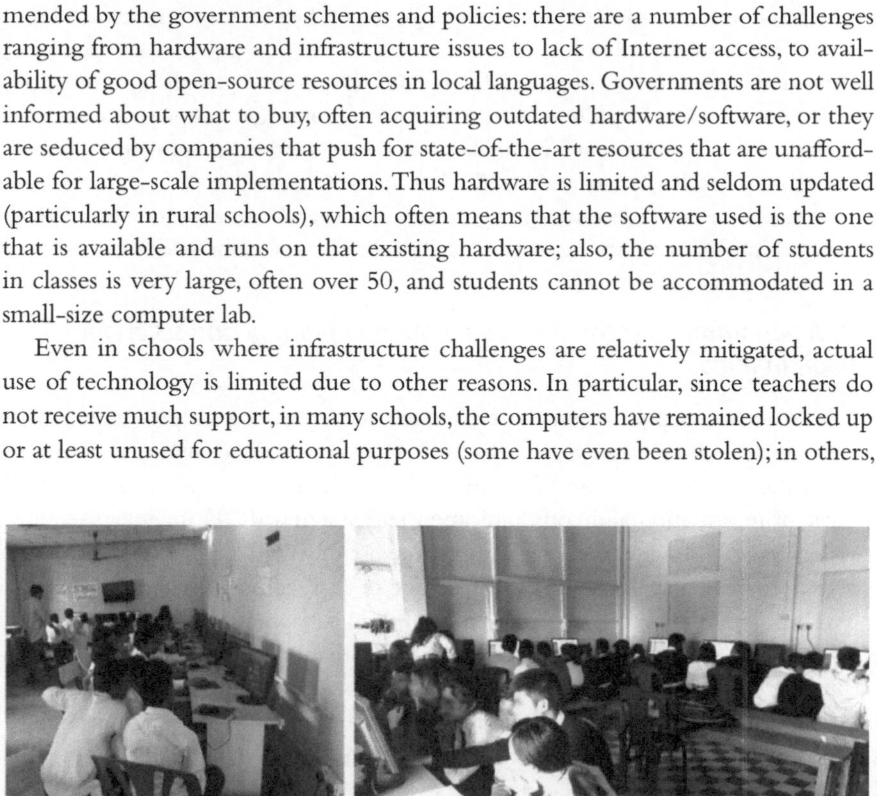

FIGURE 12.1 Students working in computer labs in Chhattisgarh and Mizoram

teachers tend to limit the use of technology to passive modes: for instance, watching educational telecasts or videos, which might have some relevance for other disciplines like language and science learning but are much less adequate for mathematical learning (Tata Trust Report, 2018). It appears that technology is either not accessible or not really being seen or used as a resource by the teacher, even when available.

In rural India, conditions are more challenging than in urban areas: there is no constant power supply and no Internet (no LAN connection in schools). However, some attempts have been made to integrate technology in such places, such as through the One Laptop Per Child (OLPC) project[6] (Shaikh, Nagarjuna, & Chandrasekharan, 2013). TurtleArt[7] (later Turtle Blocks[8] in 2013) was an integral part of OLPC and is still being widely used by young kids in India, in and out of the classroom (see Figure 12.2). And although no hardware updates have happened, the OLPC XO laptops are still used at some sites (e.g., in Khairat), even though they have become very old and slow, and the updated software versions of Turtle Blocks cannot be run. Despite the challenging situation, a study of the project in Khairat (Shaikh, Nagarjuna, & Chandrasekharan, 2013) shows it was a successful implementation with improved student engagement.

The previous OLPC example, however, is an exception. Many mathematics teachers are not confident handling technology and, therefore, shun its use with students. Also, teachers are unsure of the open-source software and tools available (e.g., LogoTurtle, GeoGebra), as they are not mapped with the existing school curriculum. And, as presented earlier, most technology efforts at the policy level are very generic with a focus on accessibility rather than the quality of learning, with little thought given to providing meaningful content or pedagogy or supporting teachers. Furthermore, they do not address subject-specific concerns, such as mathematics; even the national position paper on teaching of mathematics (NCERT, 2006) mentioned the use of technology only in passing.

In the following sections, however, some examples are given of efforts to meaningfully integrate technology in the Indian context, in particular for mathematics education.

FIGURE 12.2 Children in India using TurtleArt on the OLPC XO laptops

A.3. Addressing and coping with the challenges of technology integration for mathematics education in India

In recent times, some researchers and policymakers have focussed on identifying challenges particular to the Indian context and designing technology-based solutions to address them in order to provide more meaningful learning experiences for students, with better learning outcomes. In particular, a number of student learning apps and learning platforms have been developed in the past decade by private or other non-government agencies. In the context of mathematics education, a few noteworthy instances of these are BYJU'S[9] (a learning app), the Connected Learning Initiative (CLIx) project, and Mindspark; details of these are provided in the upcoming sections.

Many of these resources are mapped with existing curricula and have gained popularity – especially amongst students of medium-to-high-fee private schools. Most claim to improve student achievement. BYJU'S is a widely used app, focussing mainly on mathematics and science, where concepts are explained using 12- to 20-minute digital animation videos (Dwivedi, 2016). Mindspark is an artificial intelligence system for computer-aided learning (CAL), which we describe in a later section. Many others, though not backed by research, have also seen great commercial success. The relevance of these, however, is somewhat limited in terms of improving access to meaningful mathematics for the disadvantaged: their costs are prohibitive; they have high infrastructure requirements; they tend to be aimed at individual use. Furthermore, it is not clear if these apps develop a conceptual understanding of mathematics nor to what extent they enable teachers to focus on mathematical processes, such as reasoning, arguing and communicating mathematical ideas.

In the next section, we discuss one particular technology-integrated mathematics learning experiment, which tries to address, in a coordinated way, the myriad issues mentioned earlier.

A technology-enabled innovation in rural schools in East-Central India

One recent project in India, aimed at using the affordances of new technology to address the needs of students and teachers at scale is the award-winning[10] CLIx project[11]. Under this project, several learning modules[12] were developed in four subject areas, including mathematics, for government school students in four different states in India: Chhattisgarh, Mizoram, Telangana and Rajasthan. Teacher professional development and long-term support was an integral part of the project.

We focus here on instances of successful integration from a detailed study using one of the CLIx modules: Geometric Reasoning (GR)[13], developed around 2016 by a team of curriculum designers and researchers, which is aimed at junior high school students (13–15 years). About 30% of the high school mathematics curriculum in India is geometry (axiomatic Euclidean geometry), which students have difficulty with (Srinivas et al., 2016): studies show that high school students are often at an initial level of geometric thinking, while the curriculum expects them to do formal reasoning and proofs (Battista, 2007).

The GR module made use of open-source versions of existing resources, such as LogoTurtle (later Turtle Blocks JS[14]), which is a highly intuitive software for learning various geometry concepts; GeoGebra; and some other specifically designed resources, such as a game – Police Quad[15]– aimed at providing powerful and engaging mathematics learning experiences to students. These resources were customized for the CLIx module, with specific tasks aligned loosely with the existing curriculum, as it was felt that keeping them too open and exploratory would make teacher acceptance and adoption more difficult. Apart from the digital tasks, the module included hands-on activities and a student workbook. Discussions about how to relate the digital activities to classroom discussions and paper-and-pencil tasks formed an important component of teacher development and support. While for the digital activities, students had to go to the computer lab, the non-digital lessons were conducted in the classroom itself. In the sub-section that discusses the Police Quad game, we illustrate how the design element of blending digital and non-digital tasks enabled a successful integration.

In 2018, an extensive quasi-experimental learning outcomes study (TISS, 2019, pp. 34–56) was done on the implementation of the GR module in ten intervention and ten non-intervention schools in rural and semi-rural areas of Chhattisgarh, East-Central India. Several challenges were faced during implementation. In most sites, class sizes were very large (50–80 students, sometimes even 100 or more, per class), and so students couldn't be accommodated in one computer lab. To address this, the usual practice, in most of the schools, was for the teacher to divide the class into two batches, sending one batch to the lab while the other stayed in the classroom to work on workbook tasks or hands-on activities. Other challenges were limitations in the use of software: the full or latest versions of LogoTurtle (see further in the section) or of GeoGebra could not be used.

Nevertheless, the intervention school students demonstrated significantly better mathematics learning outcomes ($p < 0.001$) than their non-intervention counterparts, and the classrooms in these schools developed more student-centred norms of interaction (Srinivas et al., 2019; TISS, 2019). As the study required researchers to engage with and support teachers in the implementation for a period of six to eight weeks, it provided a clear and grounded picture of both the affordances and the challenges of integrating technology in the context of rural schools.

Addressing the challenges of technology implementation in one classroom

Along with the other intervention schools in the study, the GR module was implemented in a class of about 50 students of grade 9 (age 13–15) in a school code-named MIS10. We present this case to illustrate how technology was integrated to address some of the persistent challenges of mathematics learning and how the challenges of technology integration itself were handled.

As is the case with most government schools in India, the MIS10 school had no computers in the classrooms; it had a single computer lab to be used by the entire school, with ten old usable computers, served by two servers of 4 GB RAM each (thus each computer terminal had only 800 MB memory distribution). Moreover, the number of functional computers varied from day to day with, at most, eight working on any given day. The computer lab was spacious, but a large section of the room was unusable because of a leaking roof; thus, the computers were kept close to each other, with little space to accommodate adequately two students per computer. However, due to the class size, even after batching, often three to four students had to be seated at one computer in a somewhat cramped state.

There were software limitations because of the old hardware. For example, the CLIx team tried using the latest version of the LogoTurtle, but RAM was insufficient, so a customized version was used with only three basic commands available: forward, left and right. Using this version of LogoTurtle, students explored shapes and created designs (see Figure 12.3). Some interesting discussions were facilitated with the help of the researchers – for instance, the steps necessary to construct a triangle or pentagon.

While the technological challenges were difficult, the main challenge faced in implementation of the module was that most teachers were not able to imaginatively exploit their pedagogic affordances of LogoTurtle or GeoGebra. To support teachers with the implementation of the GR module, a workshop was held in which the researchers' team discussed with the teachers specific pedagogical content knowledge (PCK), (Ball, Thames, & Phelps, 2008) to figure out ways to address issues that emerged in class. For instance, in one classroom, a pair of students used 270° and 450° turns to construct a square instead of the usual 90°; this led to an unanticipated rich discussion about why these angles produced squares or what other angles might make squares. In this way, these regular interactions with teachers provided learning opportunities for supporting the development of PCK and deeper understandings that arose through the students' work that could have otherwise been missed.

In addition to making use of existing resources such as LogoTurtle and GeoGebra, the CLIx project also designed newer, more targeted resources in an attempt to target the specific learning challenges faced by the students, as presented next.

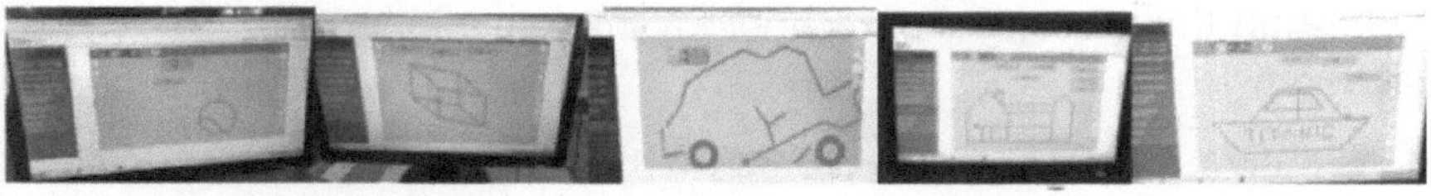

FIGURE 12.3 Students' designs using LogoTurtle

Using a learning game – Police Quad

One resource developed at CLIx to especially address the challenges of high school geometry learning (Srinivas et al., 2016) was the Police Quad game. The design decision to use a game was based on the desire to provide a safe, motivating learning environment in the classroom where students could explore and learn from their mistakes. Like other activities in the module, the game tries to help students move to progressively higher levels of geometric reasoning by doing the "spadework prior to deduction" (Shaughnessy & Burger, 1985). To make the module feasible to implement, the game was designed as blended – combining both paper and digital activities (see Figure 12.4 to Figure 12.7).

Srinivas et al. (2016) identified many misconceptions that students have about properties of 2D geometric shapes and the relationship between shape classes. These were incorporated within the game in order to create cognitive conflict (Tall & Vinner, 1981) in students while playing. This necessitated discussion and debate amongst peers in order to reach a consensus so as to not lose points and move ahead in the game. Since two to three students were playing the game together at each terminal, their disagreements often triggered meaningful class discussions. When unable to resolve through discussion, students proactively sought help from other peers or the teacher. One such episode from MIS10 is described by Srinivas et al. (2019) and illustrates how the teacher facilitated a classroom discussion to address a particularly stubborn misconception related to parallel lines. Several such episodes were observed during the study, which revealed that intervention schools spent close to 60% of class time on mathematical discussions, in contrast with non-intervention classrooms, which were dominated by teacher lecture (TISS, 2019).

FIGURE 12.4 Students trying paper prototypes of the Police Quad Game

FIGURE 12.5 Students testing digital prototypes of the Police Quad Game (source: https://www.youtube.com/watch?v=wAkoKk5_psQ)

FIGURE 12.6 Students playing the Police Quad game with teachers' facilitation in MIS09

Overall, the intervention schools showed a major change in the classroom dynamics, with students engaging actively in discourse rather than being passive listeners, as in the traditional classes. The study also reported enhanced motivation, even amongst previously disaffected students (TISS, 2019). At least two of the ten participating teachers specifically reported that a number of students generally uninterested in mathematics class had shown increased interest through the game. Also, some students with low confidence came up to the board and provided explanations.

FIGURE 12.7 A discussion in the computer lab based on the Police Quad game in MIS10

An AI-based adaptive learning programme to address individual student needs

Another digital resource that came about from the practical challenges of teaching math to classes with large PTRs is an intelligent tutoring system called Mindspark.[16] Mindspark is an artificial intelligence system for computer-aided learning, to improve mathematics and English skills. It complements classroom teaching with lab sessions in which students move at their own pace, along individualized learning paths, based on their responses. At each step, assessment data of the student is also collected, and the teacher has a dashboard to view real-time (or past) data on individual students, identify error patterns for the class, etc., for making data-backed decisions.

However, Mindspark had a major challenge in implementation – it required a 1:1 computer-to-student ratio. Thus it was used initially only in private city schools with the needed infrastructure. But more recently, experiments tried to make it accessible to government school students, such as through the Mindspark Centres: community centres set up in three low-income slum areas in Delhi where students came in batches to work on Mindspark as a part of an after-school program. The intervention included daily 45-minute sessions of personalized one-on-one, computer-aided instruction, along with 45 minutes of additional academic support by an instructor. A recent study (Muralidharan, Singh, & Ganimian, 2019) reported 0.37σ higher scores in math for middle school students enrolled for four to five months in this program compared to their peers.

Currently, Mindspark is being used on a trial basis in 180 government schools in six different states of India. In some of these schools, a different model is being tried to overcome the infrastructure limitation: students are paired, and the unit of tracking learning is the pair instead of individual students. Initial indications are that this, in fact, encourages mathematical discussion amongst peers, as in the case of the Police Quad game.

A.4. Some reflections from the experiences on technology integration in India

How technology integration impacts classroom dynamics

It may be important to examine how far the changes observed during the CLIx implementation study – creating a safe space for students to engage in meaningful mathematical discussions – would have been possible without the use of technology. Keeping in mind the ground-level challenges to technology-enabled learning, alternatives to digital activities were indeed planned and used more than once during implementation. However, such alternatives did not provide richer settings for learning or discussion than those afforded by the technology-based activities. One reason for this could be that using technology reduces the cognitive load of otherwise doing the same task on a non-digital platform, where one needs to keep track of variables that are more easily manipulated on a digital platform. The digital mode also reduces the chances of negative or censorious teacher responses to students' errors, thereby encouraging students to persevere more and use metacognitive strategies and/or peer discussion to learn from their mistakes (Srinivas et al., 2016). Moreover, digital implementation frees up teachers' time from the routine assessment of student knowledge so that they can focus on more important actions, such as facilitating class discussions (Srinivas et al., 2016). This motivated teachers to use technology and see it as an ally, which might be a critical factor in terms of sustaining teacher effort at technology integration.

Overall, the CLIx project tried a holistic approach to technology integration – with coordinated efforts that included designing tasks to encourage collaboration and discussion, as well as aspects of teacher development and sustained support. Many structural challenges were, by design, turned to opportunities for better learning (e.g., playing a single-player game in groups, which prodded students to discuss more and justify their thinking to others).

However, there are several challenges too: implementing such efforts at scale is a costly proposition both in terms of finances and human effort – but perhaps still worthwhile. Adler (2000) has argued that whether and to what extent any teaching-learning resource is a resource depends on *its use in context* by the teacher. We argue in favour of adopting this perspective in terms of policy and use of technology: one strategy is to visualize at the selection and planning stage – i.e., *before* implementation – how a technological tool could and would be used in the classroom. Especially in the case of the sample CLIx module, through the discussion of the coordinated design, implementation, and teacher development aspects, we elucidated how these factors needed to work in tandem for using technology as a meaningful resource in improving mathematics learning.

Important insights about technology integration

- *Technology integration actions need to be cognizant of the situation and context of the learner*: There are a multitude of challenges faced by the learner – from the specificity of the learning challenges to infrastructural and other situational issues – and all these aspects need to be factored in and planned for. For example, there is a need to strategize how to accommodate large classes, come up with customized or lighter versions of software and include blended programmes and non-digital substitutes of digital activities. For instance, during implementation in Chhattisgarh, one of the schools had no electrical power during the entire period of implementation, but the teacher, with support from the CLIx team, carried out a non-digital version of the Police Quad game.
- *Successful technology integration cannot happen without teacher acceptance and adoption*: This can be challenging as often teachers don't realize how to meaningfully adopt technological tools, especially those that have general purposes (e.g., Logo, GeoGebra), to teach a particular mathematics content or to connect them with the curriculum. If they use digital tools at all, they tend to prefer the less interactive options, as the system biases them towards rote learning. Hence any new technology implementation needs to be accompanied by a robust teacher professional development programme. Moreover, technologies that address some real challenges faced by the teachers in the field are more likely to see sustained adoption. A few specific examples of how the aforementioned programmes address some of these challenges are as follows:

  - In order to deal with the high PTR and teacher workload, Mindspark uses AI for learning and assessment, providing detailed feedback to the teacher. This also helps address issues of workload. The CLIx modules are designed keeping batching of students in mind to address the same PTR issue.
  - The CLIx module also tried to address the teachers' concern about covering prescribed curricular content while trying to bring in newer pedagogic practices. Instead of using the multiple affordances of GeoGebra, the CLIx module simply tried to encourage teachers to use it to get students to explore and predict properties of geometric figures (e.g., quadrilaterals). In that way, it introduced interactive classroom practices while allowing the teacher to stay close to the prescribed school curriculum.
- *There is a need to explore the scope of converting challenges to opportunities*: One good example of this was seen while the students played Police Quad: when two or more students teamed up to play the (single-player) game at a single terminal,

they collaborated naturally (except when teammates didn't get along) and engaged in constructive discourse. Hence the infrastructural challenge became advantageous from the student learning perspective.

B. The case of Mexico

As already stated, the context of Mexico, which has approximately 30 million registered pre-university school students,[17] has many common aspects with that of India in terms of diversity and infrastructure issues, as we will now discuss.

B.1. A panorama of technological integration in Mexico in government primary and secondary schools

In 1968, Mexico began making use of the then-newer technologies (i.e., television) by creating a distance education program, Telesecundaria, aimed at lower secondary students in rural areas without access to regular schools: through that school system – which still exists – learning is mediated by a teacher-promoter for all subject areas and structured through learning and content guides, television programs and, since 2006, CDs containing interactive and other resources.

In terms of computer integration into schools, in 1986, a first project, MicroSEP, was launched by the Mexican Ministry of Education (SEP) to integrate computers in schools (Ávila-Solis, 2010). Thousands of computers of the ministry's own design, which many considered by that time obsolete, were sent to schools with some accompanying materials but with practically no teacher training at a time when technologies were not ubiquitous, and teachers had no idea how to use them. Most computers were left unboxed, and the project was considered a failure.

A decade later, as described in Sacristán and Rojano (2009) and Trouche et al. (2013), between 1997 and 2007, the SEP launched two initiatives: the Teaching Mathematics with Technology (EMAT) program and Enciclomedia:

> *EMAT*, which began in 1997 (together with parallel sciences programs— *ECIT-ECAMM*) aimed to incorporate technologies in middle schools (for students from 12 to 15 years) in order to transform educational practices from the traditional teacher-to-student, top-down approach towards student-centred, exploratory, bottom-up practices. An international team of mathematics education researchers designed a constructivist, pedagogical model and activities. Universal open tools (that allowed different objectives) were preferred, such as spreadsheets, dynamic geometry (*Cabri-Géomètre*), the TI-92 algebraic calculator and, later, Logo. Emphasis was put on changes in the classroom structure, on collaborative exploratory work, and on a teaching model based on mediation and guidance […]. *EMAT* was designed to be implemented gradually – beginning with eight schools in 1997, and gradually expanding over the course of several years – so that adjustments and support could be provided, and the quality of teacher education and implementation in classrooms would be optimized.

> Though the implementation in schools was not as straightforward as planned (for example, preservice and inservice education were limited in scope [...]) – *EMAT* was groundbreaking in the ways it opened doors to integrate technologies in schools. Its use was recommended in the official national mathematics curriculum, and [had] extended beyond the originally-conceived policies. Some teachers who [worked] with *EMAT* over many years, [were] able to integrate the use of diverse tools and develop their own long-term projects.
>
> (Trouche et al., 2013; p. 760–761)

At the same time as EMAT, a new pedagogical model for the Telesecundaria schools was developed, where different types of resources were produced and articulated (printed textbooks with organized didactical sequences of activities, videos, interactive materials and the teaching guides for each subject area) and included in CDs – these resources and materials are still in use. In terms of Enciclomedia,

> *Enciclomedia* was the result of an ambitious political decision, made in 2004, to implement digitalized versions of official Grade 5 and Grade 6 textbooks in all subjects in all primary-schools in Mexico. It included accompanying digital resources and interactive whiteboards. For this project, a huge number of ad hoc interactive resources (applets) were produced in a very short time. However, the use of open universal tools, such as those used in EMAT, did not occur (Rojano, 2011). [...] Despite some successes, the haste with which *Enciclomedia* was implemented resulted in shortcomings. Rojano (2011) explained that there was an obvious jump from resource availability to its use in the classroom, with many teachers not ready to experiment and appropriate the tools.
>
> (Trouche et al., 2013; p. 761–762)

Enciclomedia aimed to provide the more than 180,000 government primary school classrooms of grade 5 and grade 6 (10- to 12-year-old students) with a smartboard, a projector, a computer, a printer and CDs with all the interactive resources. Before its cancellation, 147,000 classrooms were equipped, and over 820,000 CD kits were claimed to have been distributed (SEP, 2012). Apparently, even schools without electricity were equipped, together with some portable power sources, although anecdotal reports state that these power sources did not work.

With government changes in 2007, government support for both EMAT and Enciclomedia was discontinued. The Enciclomedia's official website disappeared, and thus it became more and more difficult to continue the use of the mathematics interactive resources (although these are still available in the distributed CDs). The smartboards were not maintained and though many are still present in primary school classrooms, they only work as projecting screens.

Since then, there haven't been any really tangible large-scale initiatives for technology integration in terms of curriculum or task implementations, neither in general nor, much less, for mathematics education.

Between 2007 and 2012, the government tried to implement a program called *Habilidades Digitales para Todos* or HDT (Digital Abilities for All) with a very different approach from those of past projects: it aimed to provide students with laptops and connect schools and teachers through "telematics" classrooms. But the results of the program are ambiguous, as stated in official evaluation documents, with only 14% of the target population covered (Arce-Torres & Guerrero, 2012). In 2012–2018, another initiative, MiCompu.Mx, was launched by SEP for providing government primary school children in grades 5 and 6, with 240,000 laptops and 1.7 million tablets; the hardware quickly failed or wasn't properly taken care of, so the project was unsuccessful (Díaz-Barriga Arceo, 2014). Both HDT and MiCompu.Mx, like MicroSEP in the 1980s, focussed on hardware, with practically no attached curriculum tasks, nor proper teacher training, and both were deemed failures. Also, Enciclomedia, though it was accompanied by some good resources for task implementations (specially for mathematics), had a huge hardware focus.

Thus, in practical terms, since before 2007, there has been no teacher training nor development of educational materials. In 2016, a big educational reform was implemented, with new textbooks in 2017 and compulsory teacher assessment meant to be a formative evaluation (which was suspended in 2019), but these did not include technology (except for some examples in some textbooks). What little technology is used nowadays in state schools either dates from the EMAT or Enciclomedia days, or it comes from the teachers' (or small local projects) own initiatives. In fact, for the 2017 reform, some examples linked with the mathematics in the textbooks were directly taken from the old EMAT materials, but that is as far as the use of technology has gotten since 2006.

Furthermore, for upper secondary school, there have been no programs of teacher training ever implemented on large scales. As reported in de Miranda Quintero and Sacristán (2016), as well as in Salinas, Sacristán and Trouche (2018), technology use at that level is much less than expected, with many task "implementations" for mathematics restricted to presentations (PowerPoint, YouTube videos or screen projections).

B.2. Some identified issues in technological implementations in Mexico

There have been several studies in Mexico researching technological integration for mathematics education in the country. The results of these point to several issues that make this integration challenging. For example, Sacristán (2013) defined two aspects: (i) the teacher's knowledge in terms of mathematics and of technological familiarity and (ii) implementation issues: lack of access to digital resources, lack of time, lack of confidence and scarce or discontinuous use. Related to those aspects, we briefly discuss the present situation in terms of two types of issues:

- *Issues of the teachers' professional knowledge and lack of professional development:* Sacristán, Parada and Miranda (2011) reported how Mexican middle school mathematics teachers lacked general computer-use knowledge and had difficulties using the Internet (lack of technology literacy). But nearly a decade later, thanks to how technology has permeated society, teachers have become much more knowledgeable in terms of the general use of computers and the Internet in their own lives (see the following section on access), though not necessarily for their mathematics teaching practice. We consider the lack of focus within the professional development (PD) of teachers to develop their mathematical pedagogical technology knowledge (Thomas & Palmer, 2014) alongside insufficient opportunities for sustained professional development (PD) as the main issues affecting adequate technological integration for mathematics education.

More recently, Salinas, Sacristán and Trouche (2018) found that even a supposedly well-trained high -school teacher in a high-level institution in Mexico City used GeoGebra simply as a projection tool for static mathematics (whereas the students took out, in hiding, tablets or smartphones to do more than the teacher).

– *Issues of access:* Nowadays, teachers (as well as many, though not all, students in urban areas) tend to have good access to computers, smartphones and the Internet at home in Mexico: in a recent 2017 study in a semi-rural and rural area of the state of Oaxaca in Mexico, 95.5% of a sample of 69 primary school teachers said they had computers at home, and 83.5% said they had Internet access at home; however, this is not the case in schools (in spite of programs like HDT). Sacristán, Parada and Miranda (2011) reported the difficulties, not only of primary and middle school teachers but also of high -school teachers in having access to technologies in the classroom (no computers in classrooms; if there are any, they are old).

In the more recent 2017 Oaxaca study mentioned earlier, out of ten schools visited, only four were equipped with computers for the students and supposedly had Internet access (but mostly unusable due to its bad quality). So teachers use their own personal laptops. In that sample, there were two extreme cases: in one school, there was a ratio of one computer per 170 students; in the one with the best ratio, it was one computer per 10.6 students. Sixteen percent (15) of computers dated from 2006, from Enciclomedia, running very slow (with the Windows Server operating system) and having issues running the pre-loaded software. The rest of the computers are newer, donated or from HDT, or they are the teachers' own personal laptops. Almost half the projectors dated from the Enciclomedia program and don't work very well.

In 2019, we visited a couple of these schools. In one of these schools, there was what is called in Mexican schools a "media room" (i.e., a computer lab; see Figure 12.8). In the other, a multigrade rural primary school, there was only the Enciclomedia 2005 hardware: a computer and the non-working smartboard (used as a projecting screen; see Figure 12.9); the teacher brought her own laptop and lent it to the children (see next section); there were also a couple of very old computers that had just been donated but that did not work.

FIGURE 12.8 A computer lab at a government primary school in the state of Oaxaca, Mexico, in 2019

FIGURE 12.9 Students in 2019 working with Logo (MSWLogo) at a rural multigrade primary school in the state of Oaxaca, Mexico, using the 2005 Enciclomedia hardware

B.3. How are teachers and schools in Mexico overcoming the difficulties: A story of a rural multigrade primary school

As mentioned previously, due to the infrastructure limitations, teachers have to use what is available: old hardware and software, as well as lend their own personal laptops. Fifteen years ago, we saw something similar in Yucatan: a teacher and all her middle school students, working with a black-and-white version of Logo on one single computer, with no hard drive to save anything and yet doing incredible computer programming and projects.

But even if they have access to some technology, most teachers still lack training and confidence and therefore hardly use it for mathematics teaching. In order to address this, a concerned teacher organized a professional development course for fellow teachers (see Enríquez & Sacristán, 2019) focussing on universal tools, such as GeoGebra and Logo, which can run on old equipment. Some of the teachers participating in the course have been putting into practice what they learn.

Using technology to teach mathematics in a multigrade rural school

In 2019, we observed one of these teachers, integrating, for the first time, GeoGebra and Logo into her math teaching, in the multigrade (grades 1–6) rural primary school in the region of Oaxaca, mentioned in the previous section. The school had zero connectivity (not even a cell phone signal) and only the old Enciclomedia computer. First, there was a group session using GeoGebra (Classic; projected). In blended tasks, children drew figures (e.g., polygons) on paper and then took turns using the one computer available to do scaled figures in GeoGebra where they had to pay attention to the scaled values. In the following session, the teacher used her own, and two borrowed laptops, so that her students could work with Logo (the MSWLogo[18] version dating from EMAT program (i.e., from 2004). Children drew polygons and figures on paper, played turtle in the courtyard and then worked (in groups of four per computer, taking turns on the keyboard) with the very basics of Logo's turtle geometry, experimenting to reproduce specific figures and then their own personal Logo projects (see Figure 12.9). The students (and the teacher) were very motivated by the experience.

The teacher explained that it was a challenge using the technological resources: in her 15 years of teaching practice, she had never used a computer for mathematics; also, some of the children had never touched a computer. She explained that encouraging her students to use the computer was not easy because she also lacked confidence in using the technology. But the outcome was very positive: she stated that using the computer and the Logo and GeoGebra resources gave her confidence and a new outlook on how to teach mathematics in which children could explore, make mistakes, do and undo, etc. She also explained that it turned her teaching upside down: instead of explaining concepts and procedures followed by examples, the class had started with experiences and problem solving (with Logo) that then led to the concepts and procedures. Thus the experience was very enriching (not only in class with the children but also when she shared it with the other teachers in the PD course). Her experience mirrors many experiences that were documented during the EMAT program (see Sacristán & Rojano, 2009).

Discussion: Issues regarding technological integration for mathematics education in developing countries

As exemplified earlier, there are many issues that are common and particular to developing countries. In both Mexico and India, there is much socio-economic diversity, and most government schools lack digital resources (infrastructure), teachers' professional development and financial resources. Assude, Buteau and Forgasz (2010) defined four types of interrelated factors that influence the integration of digital technologies in the mathematics curriculum: (i) the social, political, economic and cultural level; (ii) the mathematical and epistemological level; (iii) the school and institutional level; and (iv) the classroom and didactical level. They identified similar factors across many studies, spanning more than a decade, from different national and international settings (including those from the Mexican EMAT program) that encouraged or inhibited mathematics teachers' use of technology for mathematics learning, pointing out,

> Access to technology was a notable factor that served to encourage many mathematics teachers and also appeared to act as a barrier to others. It seems, too, that institutional or didactical factors – access to hardware, software issues, professional development needs, technical support and resources – appear to outweigh personal factors, such as confidence, in preventing teachers from using technology in their mathematics teaching.
>
> (Assude, Buteau & Forgasz, 2010; p. 416)

In the examples from India and Mexico discussed in this chapter, although some dealt explicitly with mathematical implementations, it seems that what can be done mathematically depends firstly on the availability and management of infrastructure and material resources and then on teacher knowledge and competencies. Also, of course, policies affect these aspects.

We thus identify issues which affect technological integration for mathematics education.

Issues of technology availability and access in schools

First, at the classroom, school and social levels there are *issues of technology availability and access in schools*, which highlight the digital divides. These divides are of different types – e.g., related to social and economic strata, geographical, urban versus rural, private versus government (state) schools, as well as generational – teachers versus students– and digital competencies in addition to the divide between what is available and used by students outside of schools in the digital age (e.g., mobile technologies, gaming) and what is offered in schools (as discussed by Sutherland, Robertson, & John, 2008).

In terms of the latter (the resources offered in schools), in the cases of India and Mexico, we see many limitations. First, there is limited hardware that contrasts greatly with what is used outside of schools, in spite of massive government efforts to equip schools (but with a lack of emphasis on professional development). The lack of computers in schools was also made evident in a report from Ghana where high school students are taught about ICT using a chalkboard (i.e., without computer access; Sacristán et al., 2018).

And whereas in the developed world older software and hardware tends to be replaced or discarded (sometimes because of the push of tech companies for constant renewal), in developing countries, for technical and financial reasons, teachers and schools often need to rely on those older (yet often simpler and more straightforward) technologies, as illustrated in the previous examples. That constant push for technology updates is problematic in the developing countries context: in that sense, it is ironic that even an environment as simple as Logo, which has had complete usable versions for PCs requiring very little RAM, at least since the 1980s, in the case of LogoTurtle, it could not be run in full in India. (It is for that reason that when choosing a version of Logo for EMAT in the early 2000s, a straightforward yet powerful open-source version was chosen – in spite of some international experts suggesting more "state-of-the-art" versions, and that version is still being used).

This leads us to discuss more on the types of software resources used and how they are used. As we saw in the previous examples, for obvious financial limitations, universal freeware tools are preferred, such as dynamic geometry (e.g., GeoGebra, also for its multi-usability), spreadsheets, videos or some interactive tools found on the Internet (or from the Enciclomedia CDs), and sometimes Logo, but all of these not necessarily with up-to-date versions. In general, we argue in favour of these universal free open tools.

In fact, while in many parts Logo has been put aside in favour of newer trends (e.g., Scratch), we cannot underestimate its usefulness in non-connected classrooms, as in the examples given, and for its general value for mathematical learning. And even if Logo and GeoGebra are used with limitations, in some cases, this can be quite rich and provide important pedagogical affordances by allowing students to focus on the mathematical tasks at hand. In fact, in the modern quest for ever-increasing sophistication, we often forget that simplicity often leads to richer mathematical experiences or that old doesn't mean obsolete.

Nevertheless, it is our perception that, in general, the affordances of technology are not harnessed for mathematics teaching and learning; this may be due in part to the issues of access but also due to a lack of teacher's professional development: as was pointed out by Julie et al. (2010), some of the top software used for mathematics teaching in Latin America was word processing tools and PowerPoint, so there was a dominance in the use of technologies for purposes of information and communication as is still the case. Thus if and when mathematics teachers do use technology in classrooms, it is more for presentation with either those office tools, through videos downloaded from the Internet or with math software (e.g., GeoGebra) but only for demonstration (e.g., of a graph, as in the case presented by Salinas, Sacristán & Trouche, 2018) often ignoring its affordances.

In fact, in countries like India and Mexico, mathematics tends to still be taught mainly using paper textbooks with mainly instructionist lecture modes. Only motivated teachers, or those who were fortune to participate in some PD program (e.g., from EMAT), try to integrate some technology more meaningfully, but there are difficulties fitting the use of digital resources with the curriculum which doesn't explicitly include how to use digital technologies (although in Mexico, for the new textbooks of 2018, examples taken from EMAT were used).

There is thus a need to change school and institutional cultures, which is what the Indian Police Quad game or the Mexican EMAT program attempted to do.

Going back to the issues of access, there are other aspects that are taken for granted in developed countries, such as electricity and connectivity. As we saw in the India section, in one implementation of the Police Quad game, there was no electrical power, so a non-digital version of the game had to be used. In terms of connectivity, neither Indian nor Mexican government schools have good Internet connections, if any at all (despite government efforts to change that), and thus they cannot rely on online materials, software updates, etc. And although children in urban areas have more access to the Internet, this is not the case in rural areas of developing countries (where some children have not even touched a computer, as we saw in the rural school in the Oaxaca region). In that respect, a 2017 UNICEF report claimed,

> Connectivity can be a game changer for some of the world's most marginalized children, helping them fulfil their potential and break intergenerational cycles of poverty. [...]
>
> But digital access is becoming the new dividing line, as millions of the children who could most benefit from digital technology are missing out.
>
> About 29 per cent of youth worldwide – around 346 million individuals – are not online. [...]
>
> There is a digital gender gap as well. Globally, 12 per cent more men than women used the internet in 2017. In India, less than one third of internet users are female.
>
> (UNICEF, 2017; p. 2)

Another issue in government schools in countries like India and Mexico are the large numbers of students in classes. For India, we already discussed numbers that can reach 100 students in a classroom. In Mexico, the PTR averages 27 students per teacher in primary school and 34 for secondary level (OECD, 2017). With such large classes, the availability of computers for all students, with the existing limited resources and infrastructure, is out of the question, as we have already discussed. So there is a need to strategize, such as by having several students share one computer, making batches of students attend the computer lab and planning a blended design, combining digital and non-digital tasks. Furthermore, teachers have difficulty guiding such large numbers when they work with computers. This has as one consequence: a directed or demonstrative use of technology (e.g., as in the GeoGebra case discussed earlier) that prevents harnessing its affordances for changing mathematical practices.

Issues of policy regarding technology integration

At the social and political levels, there are *issues of policy regarding technology integration*. As we saw through the examples presented, there are often government policies implemented for the sake of integrating technology, with neither pedagogical nor mathematical considerations, etc. In those cases, the efforts are generic and do not address the subject-specific concerns, and often governments are not well informed on what to buy, are seduced by companies and even acquire outdated hardware/software.

Or, in contrast, there can be proposals that seek to use technology in education for changing mathematical and classroom practices (such as EMAT, CLIx) and that might take into account research results and be implemented with well-thought-out plans (including accompanying tasks, pedagogical models and teachers' professional development). But even in the latter cases, there are issues of differences between intended curricula and enacted curricula (for a further discussion of design fidelity, see Noss & Hoyles, 2019).

Also, many initiatives are abandoned due to political changes, as in the cases of EMAT and Enciclomedia, or cut short due to lack of support (financial or professional), leading to a lack of long-term, sustained and supported initiatives. This lack of continuity in policies in the case of Mexico, and how policy affects task and curriculum implementations related to technology, was discussed by Trouche et al. (2013) who pointed out how Mexico's policies "have increasingly shifted towards access, and away from meaningful and supportive integration for mathematical learning" (p. 762), adding that social, adaptive, practical and other factors come into play that create a gap between political will and school and teacher implementation.

The issue of professional development and/or of support for technological integration and design of tasks for mathematical learning

A related issue that needs to be addressed relates to *professional development and/or support for technological integration and design of tasks for mathematical learning.* Without these, curricular or task implementations are very difficult for teachers (e.g., mathematical tasks with technology, if any, tend to be the same as paper-and-pencil ones with "technology add-ons" or a general ICT use rather than a mathematical one, as discussed earlier), and there is a reluctance, lack of confidence and even fear in teachers to use technology, as well as lack of know-how for selecting adequate resources or for using technology in significant ways as mathematical tools (Trouche et al., 2013). Implementations without these cannot be effective but they are often put aside or carried out insufficiently. In the case of EMAT, the program was designed to include research-based teacher professional development; in practice, it was much more difficult than originally conceived due to the extensive task that it implied: having to train thousands of teachers with limited previous mathematical and pedagogical knowledge as novices in the use of technology and due to the difficulties in long-term sustainment, but this case, nevertheless, stands out in comparison to others.

Final remarks

Ten years after they were written, the words of Julie et al. (2010), with reference to the different regional experiences in Africa, Asia and Latin America, still remain valid:

> The translation of policy into practice is a much more daunting task. [...] Even under massive government implementation, there remain unequal access,

> unequal resources, and sporadic use of the digital technologies in schools. Political decisions and administrative issues also affect the implementations, the quality of the training of teachers as well as its continuity and that of the projects themselves.
>
> (Julie et al., 2010, p. 380)

On a more positive note, it is encouraging how some teachers, schools, designers and trainers find ways to use what is available (whether old and limited) to provide students with innovative and rich mathematical experiences with technology and plant some seeds for the transformation of school and mathematical practices.

What remains to be done is to find other ways in which available technologies can be harnessed for enriching mathematics education, as well as agents of change to bring more digital and social equity. In particular, more affordable and widely available technologies, like smartphones and other mobile devices, are already being explored in India for more continuous teacher support and online professional (even though it is still very difficult due to access and time issues). Such mobile devices could also be explored to compensate for the lack of available digital resources in schools (even though there is a growing tendency to keep these out of schools) and for more innovative mathematical tasks, as is already happening in other countries.

Conclusion

Through the examples of experiences in primary and secondary schools in India and Mexico, we hope to have provided a glimpse into issues that are present in developing countries regarding technology integration for mathematical teaching and learning, enriching the global perspective across cultures and countries. The information and discussion given in this chapter could also have implications in terms of both practice and policy and suggest strategies for a productive integration of technology in the mathematics curriculum, as well as for teachers' professional development, that take into account socio-economic, geographical and cultural aspects, as well as the issue of technology access.

Acknowledgement

We thank Homero Enríquez for the data he provided that was used in this chapter.

Notes

1 https://www.indiastat.com/
2 http://www.conapo.gob.mx/work/models/CONAPO/Mapa_Ind_Dem18/index_2.html
3 http://dashboard.udiseplus.gov.in/#!/StatesProfile
4 http://dashboard.udiseplus.gov.in/#!/reports
5 https://nroer.gov.in/

6 http://www.hbcse.tifr.res.in/research-development/gnowledge-lab
7 See http://turtleart.org/ and http://www.playfulinvention.com/turtleart/about.html. Initially, Turtle Art versions 89, 91, 93, 146, 195 were used in sequence in the project sites.
8 https://turtle.sugarlabs.org/
9 https://byjus.com/
10 This CLIx project, devoted to using innovations in education, was awarded the Open Education Consortium (OEC) Award for Excellence in 2019 and the UNESCO award for the use of ICT in education in 2017.
11 https://clix.tiss.edu/
12 https://clixoer.tiss.edu/home/e-library
13 https://clixoer.tiss.edu/5752ad5a2e01310a05dca583/domain/Mathematics/
14 https://clixplatform.tiss.edu/turtle/
15 https://demo-clix.tiss.edu/softwares/Tools/policequad/en/
16 https://www.mindspark.in
17 https://www.planeacion.sep.gob.mx/principalescifras/
18 See http://www.softronix.com/logo.html and http://www.matedu.cinvestav.mx/~asacristan/LogoEMAT.exe for the EMAT version.

References

Adler, J. (2000). Conceptualising Resources as a Theme for Teacher Education. *Journal of Mathematics Teacher Education*, 3(3), 205–224. https://doi.org/10.1023/A:1009903206236

Arce-Torres, I. & Guerrero, A. (2012). *Informe Final de la Evaluación de Consistencia y Resultados del Programa Habilidades Digitales para Todos, Secretaría de Educación Pública. Centro de Estudios Interdisciplinarios y Prospectiva*, Santa Fé. https://www.gob.mx/cms/uploads/attachment/file/36418/Evaluacio_n_de_Consistencia_y_Resultados.pdf.

Assude, T., Buteau, C., & Forgasz, H. (2010). *Factors influencing implementation of technology-rich mathematics curriculum and practices*. In C. Hoyles & J.-B. Lagrange (Eds.), *Mathematics Education and Technology – Rethinking the Terrain. The 17th ICMI study* (vol. 13, New ICMI Study Series, pp. 405–419). New York, NY: Springer. doi:10.1007/978-1-4419-0146-0_19.

Ávila-Solis, R. (2010). Remembranza de MicroSep. Reyes Heroles: una ventana al futuro. *Pensamiento Libre*, 1(4), 12–15. www.revistapensamientolibre.com.

Ball, D.L., Thames, M. H., & Phelps, G. (2008). Content knowledge for teaching: What makes it special? *Journal of Teacher Education*, 59(5), 389–407.

Battista, M. T. (2007). The Development of Geometric and Spatial Thinking. In Frank K. Lester Jr. (Ed.), *Second Handbook of Research on Mathematics Teaching and Learning*, 843–908.

D'Souza, R. (2019). Exploring ableism in Indian schooling through the social model of disability. *Disability & Society*, 1–6. doi:10.1080/09687599.2019.1668635.

Díaz-Barriga Arceo, F. (2014). *Las políticas TIC en los sistemas educativos de América Latina: Caso México*. Buenos Aires: UNICEF.

Dwivedi, V. (13 June 2016). Why Byju Raveendran's firm is leading in the educational technology space. *The Economic Times*.

Enríquez-Ramírez, H. & Sacristán-Rock, A. (2019). *Cómo profesores de primaria enseñan matemáticas con tecnología digital: Dos estudios de caso de una zona de Oaxaca. 5o Coloquio de Doctorado de Matemática Educativa*, Cinvestav, México. http://www.matedu.cinvestav.mx/~5toColoquiodeDoctorado/progtem5to.html.

EQFI (2015). Use of Information & Communication Technology in Secondary Schools. Research Report. New Delhi: Education Quality Foundation of India. https://niti.gov.in/sites/default/files/2019-06/Final%20Report%20of%20the%20Research%20

Study%20on%20%20Use%20of%20Information%20and%20Communication%20 Technology%20in%20Secondary%20Schools%20conducted%20by%20Education%20 Quality%20Foundation%20of%20India_New%20Delhi.pdf.

ESAG (2018). Educational Statistics at a Glance. https://mhrd.gov.in/sites/upload_files/mhrd/files/statistics-new/ESAG-2018.pdfhttps://mhrd.gov.in/sites/upload_files/mhrd/files/statistics-new/ESAG-2018.pdf.

Grant, L., & Eynon, R. (2017). Digital divides and social justice in technology-enhanced learning. In Duval E., Sharples M., Sutherland R. (Eds.) *Technology Enhanced Learning* (pp. 157–168). Cham: Springer. doi:10.1007/978-3-319-02600-8_15.

Julie, C., Leung, A., Thanh, N. C., Posadas, L. S., Sacristán, A. I. & Semenov, A. (2010). Some regional developments in access and implementation of digital technologies and ICT. In C. Hoyles & J.-B. Lagrange (Eds.), *Mathematics Education and Technology-Rethinking the Terrain* (pp. 361–383). Springer US. doi:10.1007/978-1-4419-0146-0_17.

Kasinathan, G. (2009). Computer learning programs in schools moving from BOOT models to an integrated approach. http://hdl.handle.net/10625/41805.

Khan, F. (2015). *Mathematics and Its Discontents: How Well Does Mathematics Pedagogy Serve the Children of the Poor?* Max Weber Stiftung: TRG Poverty and Education Series.

Kumar, R., & Subramaniam, K. (2013). *Elementary teacher's belief about teaching of mathematics*. In G. Nagarjuna, A. Jamakhandi, & E. M. Sam (Eds.), *Proceedings of epiSTEME-5: International Conference to Review Research on Science, Technology and Mathematics Education*, (pp. 247–254). Mumbai, India: CinnamonTeal Publishing.

MHRD (2019). Unified District Information System for Education Plus – UDISE+. Ministry of Human Resource Development, Department of School Education and Literacy, Government of India. https://seshagun.gov.in/sites/default/files/update/UDISE_Booklet.pdf.

de Miranda Quintero, M., & Sacristán, A. I. (2016). *El bachillerato en México frente a las reformas educativas*. Saarbrücken: Publicia. https://nbn-resolving.org/urn:nbn :de:101:1-201608059442

Muralidharan, K., Singh, A., & Ganimian, A. J. (2019). Disrupting education? Experimental evidence on technology-aided instruction in India. *American Economic Review*, 109(4), 1426–1460

National Education Policy (NEP) (2019). *Draft National Education Policy*. Ministry of Human Resource Development, Government of India. https://innovate.mygov.in/wp-content/uploads/2019/06/mygov15596510111.pdf.

NCERT (2000). *National Curriculum Framework for School Education*. National Council of Educational Research and Training. http://ctegujarat.org/Downloads/NCF%20 Secondary%20Education-2000.pdf.

NCERT (2005). *National Curriculum Framework*. National Council of Educational Research and Training. http://www.ncert.nic.in/rightside/links/nc_framework.html.

NCERT (2006). *Position paper – national focus group on teaching of mathematics. National Council of Educational Research and Training.*

NITI Ayog (2015). Use of Information & Communication Technology in Secondary Schools: Research Report: Study Commissioned by NITI Ayog. Government of India, New Delhi: Education Quality Foundation of India. https://niti.gov.in/sites/default/files/2019-06/Final%20Report%20of%20the%20Research%20Study%20on%20%20Use%20of%20 Information%20and%20Communication%20Technology%20in%20Secondary%20 Schools%20conducted%20by%20Education%20Quality%20Foundation%20of%20 India_New%20Delhi.pdf.

NPICTSE (2012). *National policy on information and communication technology (ict) in school education. Department of School Education and Literacy, Ministry of Human Resource Development Government of India.* https://mhrd.gov.in/sites/upload_files/mhrd/files/upload_document/revised_policy%20document%20ofICT.pdf.

Noss, R., & Hoyles, C. (2019). Microworlds, Constructionism and Mathematics. [Micromundos, Construccionismo y Matemáticas]. *Educación Matemática*, 31, 7–21. doi: 10.24844/EM3102.01. http://www.revista-educacion-matematica.org.mx/descargas/vol31/2/01_REM31-2_i.pdf.

OECD (2017). *Education at a Glance 2017: OECD Indicators*. Paris: OECD Publishing. doi: 10.1787/eag-2017-en.

Portela, M. J. O., & Pells, K. (2015). *Corporal Punishment in Schools Longitudinal Evidence from Ethiopia, India, Peru and Viet Nam*. Italy: UNICEF Office of Research – Innocenti. Discussion Paper 2015-02. https://www.unicef-irc.org/publications/pdf/CORPORAL%20PUNISHMENTfinal.pdf.

Rojano, T. (2011). Recursos multimedia y el libro de texto gratuito: entre las herramientas universales y los desarrollos ad-hoc. In R. Barriga (Ed.), *Entre paradojas: A 50 años de los libros de texto gratuitos* (pp. 627–643). Mexico: Colegio de México-SEP-Conaliteg.

Sacristán, A. I. (2013). Enjeux pour le développement de compétences pour l'intégration de ressources numériques chez des enseignants de mathématiques. In T. Karsenti, S. Collin, & G. Dumouchel (Eds.), *Actes du Colloque scientifique international sur les TIC en éducation: Bilan, enjeux actuels et perspectives futures* (pp. 289–296). http://wportfolio.wzu.edu.tw/ezfiles/0/1000/academic/37/academic_75600_1071346_11995.pdfhttp://wportfolio.wzu.edu.tw/ezfiles/0/1000/academic/37/academic_75600_1071346_11995.pdf.

Sacristán, A. I., & Rojano, T. (2009). The Mexican national programs on teaching mathematics and science with technology: The legacy of a decade of experiences of transformation of school practices and interactions. In A. Tatnall & A. Jones (Eds.), *Education and Technology for a Better World* (pp. 207–215). doi:10.1007/978-3-642-03115-1_22.

Sacristán, A. I., Kaminskienė, L., Sabin, M., & Baafi, R. A. K. (2018). *WG7: Constructionism in Upper Secondary and Tertiary Levels*. In V. Dagienė & E. Jasutė (Eds), *Constructionism, Computational Thinking and Educational Innovation: Constructionism 2018 Conference Proceedings* (pp. 925–939). Vilnius University/Lithuanian Computer Society. http://www.constructionism2018.fsf.vu.lt/proceedings.

Sacristán, A. I., Parada, S. E., Miranda, L., (2011). The problem of the digital divide for (math) teachers in developing countries. In Joubert, M., Clark-Wilson, A., & McCabe, M. (Eds) *ICTMT 10: Enhancing Mathematics Education Through Technology* (pp. 243–247). Portsmouth, UK: University of Chichester/University of Portsmouth.

Salinas-Hernández, U., Sacristán, A. I., & Trouche, L. (2018). *Technology integration into mathematics classrooms: case study of a high school teachers' use of GeoGebra*. In V. Gitirana, T. Miyakawa, M. Rafalska, S. Soury-Lavergne, & L. Trouche (Eds.), *Proceedings / Actes Re(s)source 2018 International Conference* (pp. 88–91). https://resources-2018.sciencesconf.org/data/Proceedings_Re_s_source_2018.pdf.

SEP (2012). *Libro Blanco: Programa "Enciclomedia" 2006–2012*. Mexico: SEP. http://www.sep.gob.mx/work/models/sep1/Resource/2959/4/images/LB%20Enciclomedia.pdf.

Shaikh, R., Nagarjuna, G., & Chandrasekharan, S. (2013). *Socialising mathematics: Collaborative, constructive and distributed learning of arithmetic using a chat application*. In *epiSTEME 5 International Conference to Review Research on Science, Technology and Mathematics Education, Conference Proceedings*, Cinnamonteal, India, 321 327).

Shaughnessy, J. M., & Burger, W. F. (1985). Spadework prior to deduction in geometry. *The Mathematics Teacher*, 78(6), 419–428.

Srinivas, S., Khanna, S., Rahaman, J., & Kumar, V. (2016). *Designing a Game-Based Learning Environment to Foster Geometric Thinking*. In *2016 IEEE Eighth International Conference on Technology for Education (T4E)* (pp. 72–79). IEEE.

Srinivas, S., Rahaman, J., Kumar, R., Bose, A. (2019). *Analysing the role of language in students' conception of parallel lines*. In Graven, M., Venkat, H., Essien, A. & Vale, P. (Eds). *Proceedings of the 43rd Conference of the International Group for the Psychology of Mathematics Education* (p. 327, Vol 3). Pretoria, South Africa: PME.

Sutherland, R., Robertson, S., & John, P. (2008). *Improving Classroom Learning with ICT*. Routledge.

Tall, D., & Vinner, S. (1981). Concept image and concept definition in mathematics with particular reference to limits and continuity. *Educational Studies in Mathematics*, 12(2), 151–169.

Tata Trusts Report (2018). ICT implementation in school education in India- a report by Tata Trusts and IT for Change, Bengaluru, India. https://itforchange.net/sites/default/files/add/Revised%20report%20-%20Cost%20benchmarking%20of%20ICT%20implementation%20in%20Indian%20school%20education%20April%2012,%202018_0.pdf.

Thomas, M. O. J., & Palmer, J. M. (2014). Teaching with Digital Technology: Obstacles and Opportunities. In A. Clark-Wilson, O. Robutti, & N. Sinclair (Eds.), *The Mathematics Teacher in the Digital Era* (vol. 2, pp. 71–89). doi: 10.1007/978-94-007-4638-1_4.

TISS (2019). Report on learning outcomes of clix student modules. Centre for Education, Innovation and Action Research, Tata Institute of Social Sciences -TISS, Mumbai & Massachusetts Institute of Technology, Cambridge, MA, USA. https://clix.tiss.edu/wp-content/uploads/2015/09/LO_Report-2019.pdf.

Trouche, L., Drijvers, P., Gueudet, G., & Sacristán, A. I. (2013). Technology-driven developments and policy implications for mathematics education. In M. A. Clements, A. J. Bishop, C. Keitel, J. Kilpatrick, & F. K. S. Leung (Eds.), *Third International Handbook of Mathematics Education* (pp. 753–789). New York: Springer. doi: 10.1007/978-1-4614-4684-2_24.

UNICEF (Ed.). (2017). *Children in a Digital World*. New York, NY: UNICEF. https://www.unicef.org/publications/index_101992.html.

Velaskar, P. (2010). Quality and inequality in Indian education. *Contemporary Education Dialogue*, 7(1), pp. 58–93. doi: 10.1177/0973184913411200.

13

ALIGNING TEACHING WITH CURRENT EXPERIENCES OF BEING, BECOMING AND BELONGING

An identity perspective on the use of digital resources

Charlotte Krog Skott,

UNIVERSITY COLLEGE COPENHAGEN, DENMARK

Giorgos Psycharis and

NATIONAL AND KAPODISTRIAN UNIVERSITY OF ATHENS, GREECE

Jeppe Skott

LINNEUS UNIVERSITY, SWEDEN

Introduction

Studies of teachers' use of digital resources (DRs) in mathematics teaching adopt different frameworks (cf. Goos et al., 2009). Some frameworks are specialized and developed specifically for studying the use of DRs or other educational resources in classroom settings. Other frameworks adopt a broader perspective and use interpretations of issues beyond the classroom to understand how practices unfold within it. In this chapter, we intend to show the possible contribution of theories of the latter type for examining the role of the teacher and the influence of a range of personal, contextual and professional factors on the integration of DRs in mathematics classrooms. To do so, we analyze data that have previously been analyzed with a theoretical framework of the former type and compare the two analyses with regard to their respective benefits when attempting to understand the complexity of teaching-learning processes that involve the use of DRs in mathematics classrooms.

We previously used the framework called *Structuring Features of Classroom Practice* (SFCP, Ruthven, Deaney, & Hennessy, 2009) to analyze teaching-learning processes in the classrooms of a Danish elementary teacher, Sofia, who is highly committed to the use of DRs (C. Skott & Psycharis, 2018). She has just participated in a professional development project, the *Innovative School (IS)*, the aim of which is to support teachers in becoming more proficient with the use of DRs and with inquiry-based teaching. We analyzed the emerging conflicts between Sofia's conventional instructional approaches and the inquiry processes intended by IS, focussing on an episode from the implementation of an IS unit on statistics called *Youngsters' Use of Digital Media*.

In this chapter, we use another framework called *Patterns of Participation (PoP)* to analyze the same episode. PoP is more in line with the second group of frameworks mentioned earlier, favouring understandings of how broader social constellations and modes of reasoning inform action and meaning-making (Skott, J., 2019). In our choice of framework, we have taken into account recent discussions of the relationship between DRs and instruction. These discussions indicate a need to consider the broader contexts in which teachers operate as they make decisions about what resources to use and how to use them (Trouche, 2019). PoP allows a social perspective on Sofia's professional identity, which we define as her professional experiences of being, becoming and belonging at her school, Hillside. As they relate to DRs, these experiences depend, for instance, on social practices at Hillside and on broader discourses on what it means to teach with DRs and the societal position of teachers in general. In fact, from a PoP-perspective, experiences are constituted in social interaction. Using PoP, it becomes important to investigate Sofia's engagement with a multitude of different practices beyond the ones promoted by IS and their possible significance for her professional experiences. Our research questions are as follows:

- How does Sofia's experiences of being, becoming and belonging at Hillside influence how DRs are used in her classroom?
- What are the potentials of social practice theories for understanding teachers' contribution to classroom practices involving DRs?

We begin by recapitulating the key characteristics of SFCP and the results of our SFCP analysis of the data on Sofia. Next, we outline aspects of PoP before presenting the methodology and the results of our PoP analyses. We end with a discussion and the contribution of our approach.

SFCP and the SFCP interpretation of Sofia's classroom

The SFCP framework synthesizes concepts from earlier studies of classrooms and teacher expertise and applies the synthesis to technology integration. SFCP, then, is a model for analyzing teacher expertise as it relates to the integration of DRs. The question that orients SFCP is "how material-cultural factors structure the functional organization of teaching expertise" (Ruthven, 2014, p. 390). SFCP intends to make it possible to analyze crucial facets of teachers' expertise as "craft", facets that are related to classroom experience rather than to formal teacher education. SFCP highlights five features that shape how teachers do or do not integrate DRs: (1) *working environment*: the physical surroundings, organization of people and resources and routines enabling lessons to flow smoothly; (2) *resource system*: the coordinated use of tools and curriculum materials for student activity and learning; (3) *activity structure*: the collection of activity formats, which frame how the teacher and students act and interact in specific parts of a lesson; (4) *curriculum script*: a loosely ordered individual system of goals, resources and actions for teaching a mathematical topic that interweaves mathematical ideas, appropriate tasks, potential emergent issues, anticipated student responses and alternative paths of action; and (5) *time economy*: the teacher's management of the time to convert it into "didactic time" for student learning.

Previously, we used SFCP to analyze if and how Sofia aligned her teaching with the aims of IS. In particular, we were interested in if and how Sofia incorporated new DRs into her resource system and how their classroom use influenced her activity formats and curriculum script. We focussed on Sofia's contributions to classroom interactions that involved DRs to interpret emerging conflicts between the activity structures related to her history of classroom teaching and the inquiry processes promoted by IS. Our analysis resulted in the somewhat surprising finding that Sofia tended to stick to her typical procedure-oriented activity format when DRs were involved but to align with IS when they were not. It seemed that the integration of DRs reinforced Sofia's procedural contributions to classroom interaction that supported the students in solving closed tasks and provided them with technical support.

As part of IS, Sofia taught the unit on *Youngsters' Use of Digital Media*. The unit exploits the affordances of different digital and non-digital resources to support students' development of conceptual understanding and a critical stance to the use of statistics. We analyzed an episode, *the calculator episode*, to which we return later in this chapter. In this episode, Sofia asks a group of students to use a calculator to check if the mean of four time periods (7 hours and 48 minutes) given in a newspaper article is correct. The group has difficulties interpreting the decimals provided by the calculator as minutes. Sofia is unsuccessful in helping them, and the episode ends with both parties being dissatisfied.

The SFCP analysis shows how Sofia's conventional activity format makes it hard for the students to engage meaningfully in the intended statistical inquiry. Regarding the resource system, Sofia chooses to use a DR familiar to the students (the calculator), bypassing DRs suggested by the unit (e.g., spreadsheets) that could promote their inquiry. When planning the implementation, Sofia draws on her existing curriculum script, built from years of experience, which focusses on using DRs for standard procedures. Furthermore, she ignores that refinements are needed in order to use DRs to support inquiry-based learning. In effect, she blocks the students' communication and their use of DRs in statistical inquiry.

Identities and the PoP framework

Mathematics education has focussed increasingly on the role of social and broader cultural and societal factors when seeking to understand instruction. This shifts the emphasis from teachers to teaching (Chazan, Herbst, & Clark, 2016; Skott, Mosvold, & Sakonidis, 2018), and explanations of teachers' acts and meaning-making tend to focus less on teacher characteristics per se, especially their knowledge and beliefs, and more on how institutional contexts and other social factors are reflexively related to and co-determiners of instructional decisions and classroom practice.

PoP is in line with this development. It was initially developed in response to the conceptual and methodological problems of research on teachers' beliefs and to develop a more social perspective on affective issues in mathematics education (e.g., J. Skott, 2015a, 2015b). It has later been used in studies of teacher identity (J. Skott, 2019). Like a range of other studies of teacher identity (e.g., Buchanan, 2015; Kang & Battey, 2017), PoP draws on the notions of practice and figured worlds in social practice theory. In this line of thinking the term practice "connotes doing [but] doing in a social and historical context that gives structure and meaning to what we do [...] practice is always social practice" (Wenger, 1998, p. 47). Figured worlds are "socially and culturally constructed realm[s] of interpretation, in which particular characters and actors are recognised, significance is assigned to certain acts, and particular outcomes are valued over others" (Holland, Skinner, Lachicotte Jr., & Cain, 1998, p. 52). Holland et al. mention caste systems in Nepal and romance at a university campus as examples of figured worlds.

However, identity studies with PoP differ from other studies of professional identity that draw on the same or similar social perspectives on human functioning (cf. J. Skott, 2018). This is so in the sense that we do not focus on how teachers move towards more comprehensive participation in one particular, pre-established practice (e.g., a professional development (PD) programme) or figured world (e.g., the reform). Instead, we recentre the individual and seek to understand what and how other previous and present practices and figured worlds play a role in the teacher's contributions to classroom interaction and how the character of and relationships amongst these practices and figured worlds change over time. To do so, we draw on the notions of self and interaction in symbolic interactionism (Blumer, 1969; Mead, 1934). The notion of self has two aspects or phases that evolve in interactions: an I and a me. When the I acts, the actor instantaneously sees herself or himself through the eyes of others; interprets their actual or envisaged reactions to her or his own act, be it physical or verbal; and adjusts the act in the process. In the terms of symbolic interactionism, the acting person takes the attitude to herself or himself of others and becomes a me.

Consider, for example, a teacher who in the process of explaining a solution to a student interprets the student's increasing physical uneasiness as an indication of mental and emotional withdrawal. Taking the student's attitude to herself, the teacher may react emotionally and adjust the explanation – for instance, by leaving out conceptual difficulties and focussing on the procedure to avoid difficulties that may further threaten the student's involvement and her own relationship with the student. In this case, "the other" is the student. However, one may also take the attitude to oneself of generalized others – that is, of a social group or community. In PoP, we interpret practices and figured worlds as possible generalized others. In a study of the professional identities of a novice teacher, Anna, for example, the focus was on how Anna's professional experiences developed over the first few years of her career (J. Skott, 2019). The character and importance of some significant others changed (e.g., the team of colleagues she worked with and the reform discourse), and she began to take the attitude to herself of other ones (e.g., the school leadership). As the dominant others changed, her professional experiences were transformed, including her relationships with the students, her position at the school and the ways in which she engaged her students in mathematics.

In the case of Sofia at Hillside, we also use PoP for identity purposes. As in other studies, we define identity as Sofia's experiences of being, becoming and belonging as they relate to the profession. This may include experiences of being (in)competent or challenged; of becoming better (or worse), for instance, at mathematics, at using DRs or at establishing productive relationships with her students; and of belonging by being recognized as a good colleague or as a good teacher by the students, the parents and the leadership. Identity is not seen as a stable personality trait but as dynamic and affectively laden. When Sofia is with her students, for instance, her experiences of assuming professional responsibility and being recognized as a competent professional may change if she at the instant sees herself from the perspective of a recent PD initiative or a dominant discourse amongst her colleagues on what it means to teach. Each of these and many other practices or figured worlds may influence Sofia's shifting experiences of assuming – or not assuming – her professional responsibilities. We use PoP to analyze how Sofia engages with other individuals and generalized others, including practices and figured worlds, as she interacts with the students and the contents. For our present purposes, we are particularly interested in if and how the use of DRs is incorporated into and transforms the dominant practices and figured worlds that orient Sofia's contributions to her interactions with the students.

Methodology

Sofia at Hillside

Sofia graduated from college in 2000. She had studied to become a teacher of mathematics, physical education, science and domestic science. However, she experienced college mathematics as similar to mathematics in high school, as it – according to her – focussed on reproductions of theorems and proofs, not on educational matters such as classroom use of DRs. Upon graduation, Sofia took up teaching at the small, rural elementary school, Hillside, which has one class at each grade level from 1 to 6. Sofia still works at Hillside and teaches the aforementioned four subjects in grades 4–6.

In 2013, Hillside merged with two other schools in the municipality and became one of three departments of a large elementary and lower secondary school. It is a top priority of the new school's leadership to develop teaching with DRs. This is in line with the educational strategy of the municipality that has invested heavily in digital equipment for all its schools. In 2013, the leadership at Hillside applied for the school's participation in IS in order to build teachers' capacity to teach with DRs.

When Hillside's application is accepted, Sofia is the only mathematics teacher who is asked by the leadership to take part in both rounds of IS, and consequently, she teaches the unit (15 lessons) *Youngsters' Use of Digital Media* twice, both times in grade 6. She is also asked to be the local leader of ten colleagues who teach this unit and another unit taught in grade 3. As the IS initiative finishes in 2015, the leadership of the school and the educational leadership of the municipality agree that the school has benefitted tremendously from its participation. The leaders, therefore, decide to continue the teaching of all five IS units. Again, Sofia is asked to lead her colleagues' teaching of the two units. Today, the teachers still teach these units, and the school continues to emphasize the significance of IS. This is so to the extent that the school's participation in the project is mentioned at the top of its home page.

Sofia considers herself digitally competent, and she is recognized as such at Hillside. She has participated in many other development projects on the use of DRs in mathematics and, as indicated earlier, Hillside supports her in doing so. This means that the case of Sofia at Hillside may be seen as a critical case (Flyvbjerg, 2011), as it allows a generic interpretation: if the potentials of IS do not materialize in this case, they may be unlikely to do so also under less favourable circumstances. In combination with PoP, the study of Sofia at Hillside becomes instrumental (Stake, 2000), as it sheds light on the more general issue of how teachers' professional experiences are co-constituted by practices and figured worlds in and beyond the school and the classroom.

Data generation

We used field notes and video to observe Sofia's classrooms and to identify differences and similarities between her normal approach to teaching and the ones she took when teaching *Youngsters' Use of Digital Media*. We observed 15 lessons before and after Sofia's teaching of the unit and 16 lessons from the unit itself, mainly from the second round in which we assumed that Sofia was more familiar with it. We transcribed 15 hours of our recordings. To understand what other practices and figured worlds may influence Sofia's use of DRs, we conducted six semi-structured one-hour interviews. Four of them were with Sofia, three during IS and one five months later; two were focus-group interviews, one with Sofia and two of her colleagues and one with the school leadership. The two latter interviews focussed on IS, especially on how Sofia and her colleague engaged in it. We transcribed all six interviews.

Data analysis

We conducted two analyses. The first was to identify practices and figured worlds that may play prominently in Sofia's use of DRs when teaching. Inspired by grounded theory (Charmaz, 2014), we analyzed the four individual interviews with Sofia line by line and constructed initial codes, such as "differentiating teaching is easier with DR" and "supervising rather than teaching". Based on these, we constructed focussed codes, for example, "using DR to meet general pedagogic concerns", which we assembled into distinct practices and figured worlds. In this way, we identified two characters from practices and figured worlds related to the teaching of mathematics: *a modern teacher* and *a digital spearhead* and a contrasting figured world that Sofia still refers to, *the traditions of mathematics teaching*. In what follows, we refer to these three as identity markers. We confirmed and consolidated these markers by comparing extracts from the two other interviews and from classroom observations.

The second analysis was to investigate if and how the three identity markers played significant roles as generalized others for Sofia when teaching. In order to make it possible to compare our SFCP analysis with the PoP analysis, we chose the calculator episode referred to earlier. We also analyzed other classroom episodes to consolidate and refine our findings.

Prominent identity markers

We now present the three identity markers identified in our first analysis and show how they play significant but different roles in Sofia's use of DRs.

Being a modern teacher

A significant character for Sofia is that of a progressive and modern teacher. There are two aspects to this: the use of DRs and elements of the reform agenda.

As for the first aspect, Sofia is particularly concerned with introducing new and technologically advanced DRs (e.g., multimedia applets). She focusses on whether they allow the students to engage in different activities and to have fun, as well as allow her to differentiate activities so as to engage all students, particularly the weaker ones. Another rationale is to ease her obligations – for instance, correcting tasks and controlling student work. In her own account, the use of DRs has not transformed her approach to teaching, but it has made certain things easier.

As for the second aspect, Sofia is inspired by constructivism, which has two implications for her. The first implication is that the students should be active with the content and responsible for their own learning. The second implication is that Sofia is reluctant to present mathematics to the whole class and to engage in whole-class discussions. She considers "that old-school teaching" (second interview) a waste of time as most of her students can or will not follow. Rather, she sees her main responsibility as being a supervisor, organizing students' independent learning.

These two aspects merge for Sofia into imaginings of a modern teacher, who uses DRs effectively within a reform agenda. We describe two ways in which she relates to *being modern*. Each of them points to ways in which she disclaims responsibilities often attributed to teachers.

Firstly, acting as a *modern teacher*, Sofia refers to the reform agenda by implying that students need to engage actively with DRs. She even expects them to learn to work with new DRs by themselves: "That is what I expect of them. When we were to learn to use GeoGebra [I told them] 'You have to learn it by yourselves'" (first interview).

Secondly, and because of leaving the responsibility for the students' learning to themselves, Sofia does not need to know everything about the DRs used in the classroom. She happily asks students about the potentials of new DRs and about how to solve tasks with them. This makes it possible for her to introduce new and technologically advanced DRs without having the responsibility of familiarizing herself with them in advance. It also means that she does not need to solve the tasks the students are to work with beforehand, irrespective of whether they require new or complicated uses of DRs. She is not worried that she is sometimes unable to help the students: "That is how it is to be a teacher" (second interview).

To summarize, Sofia's experiences of *being a modern teacher* relates to her use of new and technologically advanced DRs that frees her from responsibilities that teachers traditionally assume. By *being modern*, she distances herself from most of her colleagues and positions herself as different. At the same time and paradoxically, she claims to gain a prominent and valued position amongst them.

Relating to the traditions of mathematics teaching

Sofia also draws on approaches to school mathematics that are more traditional. This is so in two ways. First, Sofia emphasizes students' mastery of skills and tools (e.g., multiplying fractions, using a protractor) and does not acknowledge the significance of mathematical processes. She focusses on "all the basic knowledge… all the content parts of mathematics" (first interview). Though Sofia is aware of a new emphasis on processes in the national syllabus, she rejects it and trusts her own professionalism: "[it is] too fluffy in relation to how concrete the subject really is" (second interview). Student command over basic skills is an aim in its own right that requires practice: "It is important to make a lot of the same type of tasks because then you get good at it" (first interview). However, Sofia also refers to the assumption that drill and practice lead to conceptual understanding.

Second, Sofia's choice of a particular DR concerns whether it allows students to solve their normal tasks quickly and easily. It is not based on whether the DR provides students with opportunities for learning concepts and procedures more profoundly or for engaging students differently in mathematical processes. Spreadsheets, for instance, involve too many choices and do not help the students produce answers quickly. Therefore, Sofia does not use spreadsheets, although the national syllabus requires her to do so.

These two issues suggest that mathematical proficiency is a matter of solving routine tasks quickly and point to efficiency as a key concern for Sofia's choice and use of DRs. Some DRs introduce efficient methods for dealing with, for instance, geometric problems. This includes GeoGebra, as it helps students reflect and translate geometric figures more easily than when drawing on paper. Others are digital platforms offering instantaneous feedback when practising skills: "[They] are fantastic for drill and practice tasks compared with providing students with notebooks" (second interview).

To summarize, Sofia focusses on traditional mathematical virtues and on DRs that provide students with efficient ways of learning and working with the related routines. Her experiences of being valued by her colleagues and students and of being aligned with what is generally required in the profession are based on the extent to which she manages to establish teaching-learning situations that allow the students quick and easy access to the routines. Being valued and competent with regard to DRs is then a matter of the procedural efficiency of the resource, not of learning the specific content differently or more profoundly.

Being modern merges with the traditions of mathematics teaching

Being modern merges sometimes for Sofia with the *traditions of mathematics teaching*. One example is when Sofia designs her lessons. When doing this, she relates primarily to the *traditions* by basing most of her decisions about the content on a textbook that matches her emphasis on routines and using its tasks (i.e., closed questions with right answers) and what is referred to as "workshops" (i.e., repetitions of skills in new contexts). Normally, she begins a lesson by telling students which tasks to do and providing organizational details. After that, the students work for some time individually on routine tasks on a digital platform, and later, they work in groups with the textbook tasks or workshops. As a *modern* teacher, Sofia supervises groups of students and rarely presents specific content to or leads discussions with the whole class. Hence Sofia promotes a reform agenda regarding student's autonomous learning at the surface level, while she engages them in routine tasks and guides them on procedural and technical matters when she interacts with them.

Becoming a digital spearhead

As mentioned previously, using DRs is to Sofia one part of *being a modern teacher*. Gradually, she also begins to see herself as a *digital spearhead*. There are two interrelated aspects to this. One is her experience of being digitally committed and competent; the other is how she positions herself and is positioned as proficient with DRs at the school and elsewhere. As for the first, her major motivation for using DRs is her personal commitment, which has taken her beyond what could be expected of a teacher: "I have invested interest in this.... To me, this is not just a job.... If it was really just a job, I wouldn't do it!" (first interview). One indication of her commitment is that she has participated in many PD activities related to DRs in order to keep on top of recent developments. Also, she refers to discussions with her husband, who teaches mathematics at another school, about how best to use DRs.

In part, the second aspect also concerns Sofia's participation in PD, including IS. Her participation has not made her think of herself as part of a community of highly qualified and committed DR users. On the contrary, she consistently refers to her disappointment both with the other participants, who know much less than she does, and with the PD initiatives, as they do not support her in becoming better at using DRs. She always expects that "[t]here must be somebody, who knows more [than me]…who can inspire me, but there never is" (first interview). Considering herself more digitally competent than her colleagues at Hillside and beyond, Sofia distances herself from them and positions herself as a digital spearhead. This is affirmed by how she is positioned by the school leadership as key to the successful integration of DRs amongst her colleagues. She is highly valued for her digital competence and asked to participate in and be in charge of various activities, such as being the digital coach for all the schools' mathematics teachers. In the interview with the leadership, they readily acknowledge her significance for the school's integration of DRs.

In summary, Sofia consistently experiences herself as being highly digitally competent, and she positions herself and is positioned as such. Her sense of *becoming a digital spearhead* coevolves with her distancing herself from other teachers. Thus as a *digital spearhead* she experiences belonging to an imagined elite of teachers who are experts on the use of DRs in mathematics teaching.

The calculator episode

We now return to the episode described earlier in which Sofia asked a group of students to use a calculator to find the mean of four time periods. Our analysis aims to document if and how the three identity markers described in the previous section play a role as generalized others for Sofia when she interacts with the students.

The activity

In the second round of IS, Sofia teaches *Youngsters' Use of Digital Media* to a small grade 6 class with 13 students. Sofia considers these students mathematically weaker than the class she taught in the first round. The unit's first lessons are based on a newspaper article that uses inappropriate statistical methods to justify and argue that youngsters spend too much time on the media. It is a main goal that the students develop a critical stance to the use of statistics by examining the article. The students are to use spreadsheets to work on open questions – for instance, on how the journalist has presented the data and used them to calculate the mean of previously found averages of media time and how to interpret this mean (7 hours and 48 minutes).

The incident

The incident lasts 12 minutes and involves interactions between Sofia and a group of three girls who according to Sofia are amongst the mathematically strongest in the class. Even before the girls have read the questions, Sofia asks them to use a calculator to check the mean of the four averages, thus omitting the critical and difficult part of the questions. She also tells them how to transform time into minutes. Following Sofia's instructions, the girls transform each average, add the numbers and divide the sum by 4. Sofia intervenes, asking them also to divide by 60 without further explanations. During the next five minutes, they and Sofia struggle to get the calculations right. Finally, the calculator returns 7.7333, and the girls utter, "It was close.... But not right". Unsure how to interpret the result, one girl asks Sofia, but Sofia replies that she must find out herself. Instead, Sofia begins to explain how to change 0.7333 hours to minutes: "If you multiply by 60 then you get that part of an hour". However, the girls do not understand her explanation, and one girl complains, "You confuse me more than I was before". In the end, both parties seem disappointed. Sofia comments that the students do not develop understandings from the procedures: "But you do not understand; only know how to do it".

Analysis and results

In what follows, we use the three identity markers to analyze Sofia's instructional choices when planning the activity and her interactions with the students as she implements it. This allows us to develop an understanding of the role of each marker and their shifting mutual relationships. This leads to an identity interpretation of why the calculator episode develops as it does.

Identity markers when designing the activity

At the level of design, we emphasize three of Sofia's choices that relate to the identity markers, sometimes in contradictory ways. First, Sofia does not generally deal with the inquiry-based character of IS activities, and in particular, she aligns herself with the *traditions of mathematics teaching* by turning the open questions of the activity into tasks for practising procedures. This choice, however, may also relate to the students being relatively weak, prompting Sofia to draw on her obligations as a *modern teacher* and simplify the task to make it accessible to all students. In this case, elements of the first and second identity markers seem to merge.

Second, Sofia rejects the use of spreadsheets and decides to use a calculator, a tool that in this activity does not promote student inquiry. This choice relates to all three identity markers. After narrowing the open questions, a calculator is more efficient for finding the right answer, and in this sense, Sofia's choice aligns with the *traditions*. *Being modern*, Sofia normally chooses a new or technologically advanced DR and not a calculator. She compensates for this in this case by using it in combination with a multiplicity of other DRs. Also, and still as a part of *being modern*, Sofia's choice of the calculator can be seen as guided by a general concern for the weaker students, who are less challenged by a well-known DR. Finally, Sofia's decision relates to how she sees herself as more digitally competent than others. Being the only participant in IS to question its suggestion of DRs, she draws on her digital commitment and confidence – that is, becoming a *digital spearhead* – to challenge the project recommendation.

Third, and in line with *being modern*, Sofia has not prepared herself by working with the activity beforehand or considering how to facilitate her students' learning. According to her, the students are to learn by engaging with the content and the calculator themselves, and her role is one of framing and guiding their work.

Summarizing, Sofia makes three choices oriented by the identity markers when preparing the activity. *Being modern* plays the most dominant role.

Identity markers when implementing the activity

Although all of the three identity markers are at play in the incident, the balance between them changes. The interactions relate mainly to *being modern* and *the traditions* while *becoming a digital spearhead* coexists in the background. We divide the analysis into three parts.

First, it is notable that Sofia narrows the questions to such an extent that even the most mathematically strong students cannot approach them in other ways. This indicates that making questions more skill-focussed relates more to Sofia's emphasis on *traditions* than to a *modern* concern for the weaker students.

Second, Sofia expects the girls to engage actively with the content and the calculator. In line with *the traditions*, she provides explicit procedural support but aligns with *being modern* when the girls request a meaningful interpretation of the mean, and she requires them to discuss it amongst themselves. Her rejection to engage in conceptual discussions can also be seen as rooted in *the traditions*, as her assumption is that conceptual understanding develops when students work with procedures.

Third, Sofia is not prepared for the conflict that emerges with the students. The girls' unproductive struggle with the calculator and their growing frustration suggests that at least the part of *being modern* related to differentiating instruction and allowing students to have fun are challenged. As the calculator does not provide a meaningful context for the students, Sofia increasingly draws on her well-established *traditional* orientations, thus re-engaging in practices of supporting students' mastery of skills. This tension between the group's emotional responses to not understanding and Sofia's increased procedural instructions create a communicative gap suggesting a conflict amongst Sofia's identity markers. That is, featured choices such as the *modern* prompt to let students work autonomously finding their own ways to solve tasks with DRs, the *traditional* priority given to students' skill mastery and to providing procedural instructions and the *digital-spearhead* decision to do away with the recommended DRs and use a calculator instead short-circuit her interactions with the girls.

This conflict may influence Sofia's professional experiences and orient her future ways of contributing to interactions with students. We wonder whether the episode will reduce Sofia's engagement with the *traditions of mathematics teaching* or – more likely – will lead her to blame IS and the unit for the conflicting communication.

Discussion and conclusion

Our previous analysis based on the SFCP framework provided a rich descriptive account of conflicts between Sofia's dominant activity format and the innovative ones targeted by IS. For instance, it prevented students from engaging meaningfully in the intended statistical inquiry when DRs were involved.

In this chapter, we adopt a more explanatory approach. We revisit the data and use PoP to relate the emerging classroom practices to significant aspects of Sofia's experiences of being, becoming and belonging as a teacher at her school – that is, of her professional identities. Our interactive interpretation of these experiences points to the significance of social practices in and beyond the classrooms and suggests an analytical potential of a social perspective on identity for understanding teachers' contribution to classroom practices with DRs.

We identify three practices, characters and figured worlds that orient Sofia's experiences at Hillside. These identity markers are *being a modern teacher*, *relating to the traditions of mathematics teaching* and *becoming a digital spearhead*.

At the level of design, we see that all identity markers merge in Sofia's decision to use a calculator and not a spreadsheet. *Being modern*, however, dominates at this level, as Sofia disclaims responsibilities often attributed to teachers such as investigating the learning potentials of the activity and addressing students' difficulties with conceptual issues in advance. This is so even when Sofia as a *digital spearhead* decides to use a different DR than that suggested by IS. When teaching, Sofia acts as a *modern* teacher by disclaiming responsibilities for setting up whole-class sessions, which combines well with *the traditions* but poorly with IS. Thus, although Sofia claims to follow a reform agenda, as does IS, she only does it on a surface level, expecting students to work autonomously but with tasks that she has transformed into closed ones targeting skill mastery and with DRs that she has not prepared herself to use. In our interpretation, her affiliation with these dominant figured worlds prevents her from attempting to engage in the parts of IS that concern DRs as resources to promote students' statistical understandings and engagement in processes of inquiry and reasoning. This constellation of identity markers, including the recognition of her as digitally competent from the school leadership, makes Sofia very confident in her ways of using DRs.

The analysis of the incident indicates a pattern in Sofia's contributions to classroom interactions. PoP allows us to analyze this pattern and highlight how and to what extent it is oriented by the three prominent identity markers. The analysis brings to the fore how Sofia's identity fluctuates as she engages with her students while her experiences of assuming professional responsibility and of being recognized by her school and beyond as a competent teacher working efficiently with DRs evolve. Initially, Sofia draws mainly on *being modern*, assuming that the students would solve the tasks by themselves. As the incident evolves, Sofia insists on providing "efficient" explanations in line with *the traditions* but fails to support the learning of her better-performing students and to align with IS's inquiry-based aims. This experience may and may not at the instant challenge her sense of herself as being a competent teacher. In the subsequent interview, however, she explains the students' reactions with reference to the poor design of the unit, keeping intact her experience of herself as a *digital spearhead*. This may be seen as an answer to our previous question about whether Sofia will reduce her engagement with traditional mathematics or blame IS.

In the above sense, the PoP analysis offers a set of explanations for the descriptive account of Sofia's use of DRs in our previous SFCP analysis. In particular, the analysis can explain Sofia's dominant focus on procedures at the level of design and in classroom interactions when DRs are involved. More generally, PoP offers an identity perspective on teachers' roles in technology-based learning environments and on how figured worlds, characters and practices beyond the classroom may influence the integration of DRs in different ways. PoP allows a perspective on significant identity markers as they relate to different aspects of teachers' professional lives. While the SFCP analysis is useful in providing descriptive accounts of aspects of Sofia's instructional approaches, PoP offers more social explanations of the development of these aspects. This takes the analysis further and allows contextual interpretations and explanations of Sofia's agency – that is, of her acts as related to broader social enterprises beyond the immediate interaction.

The insight provided by the aforementioned findings is not limited to a single teacher. The generic case of Sofia indicates the complexity of how teachers' engagement in different practices and social worlds (in and out of the classroom) interrelates with their professional experiences as regards the integration of DRs in teaching. We expect that different, similar or other identity markers, patterns of classroom interaction and respective explanations will emerge in other less promising cases of teachers who attempt to use DRs in their mathematics classrooms.

At the level of theory, the two frameworks help us highlight different facets of Sofia's experiences with using DRs in her classroom: while SFCP provides an account of her organization of the lesson and her structure of interactions with the students, PoP offers an identity lens to analyze if, how and why the use of DRs influences the practices and figured worlds orienting her contribution to the classroom interactions. In conclusion, the present study indicates that different frameworks and theoretical approaches – including networking and double analyses – are needed to develop balanced understandings of the integration of DRs in mathematics classrooms that acknowledge the significance of teachers' professional identities.

References

Blumer, H. (1969). *Symbolic Interactionism: Perspective and Method*. USA, Berkeley: University of Los Angeles Press.

Buchanan, R. (2015). Teacher identity and agency in an era of accountability. *Teachers and Teaching*, 21(6), 700–719. doi:10.1080/13540602.2015.1044329.

Charmaz, K. (2014). *Constructing Grounded Theory*. London, England: SAGE.

Chazan, D., Herbst, P. G., & Clark, L. M. (2016). Research on the teaching of mathematics: A call to theorize the role of society and schooling in mathematics instruction. In D. H. Gitomer & C. A. Bell (Eds.), *Handbook of Research on Teaching* (5th ed., pp. 1039–1097). Washington, DC: AERA.

Flyvbjerg, B. (2011). Case study. In N. K. Denzin & Y. Lincoln (Eds.), *The Sage Handbook of Qualitative Research* (4th ed., pp. 301–316). Thousand Oaks, CA: SAGE Publications.

Goos, M. et al. (2009). Teachers and teaching: Theoretical perspectives and issues concerning classroom implementation. In C. Hoyles & J.-B. Lagrange (Eds.), *Mathematics Education and Technology-Rethinking the Terrain. The 17th ICMI Study* (pp. 311–328). New York, USA: Springer.

Holland, D., Skinner, D., Lachicotte Jr., W., & Cain, C. (1998). *Identity and Agency in Cultural Worlds*. Cambridge, MA: Harvard University Press.

Kang, H. J., & Battey, D. (2017). Contextualizing the identity development of preservice elementary mathematics teachers in methods courses and mentor teacher support. *Journal of Education and Development*, 1(1), 24–36. doi:10.20849/jed.v1i1.248.

Mead, G. H. (1934). *Mind, Self, and Society from the Standpoint of a Social Behaviorist*. USA, Chicago: University of Chicago.

Ruthven, K. (2014). Frameworks for analysing the expertise that underpins successful integration of digital technologies into everyday teaching practice. In A. Clark-Wilson, O. Robutti, & N. Sinclair (Eds.), *The Mathematics Teacher in the Digital Era* (pp. 373–393). Dordrecht: Springer Science & Business Media. https://doi.org/10.1007/978-94-007-4638-1_16.

Ruthven, K., Deaney, R., & Hennessy, S. (2009). Using graphing software to teach about algebraic forms: A study of technology-supported practice in secondary-school mathematics. *Educational Studies in Mathematics*, 71(3), 279–297.

Skott, C. K., & Psycharis, G. (2018). *Studying the use of digital resources in mathematics classrooms: A deeper focus on the reasons underlying teachers' choices*. In H.-G. Weigand, A. Clark-Wilson, A. Donevska-Todorova, E. Faggiano, & J. Trgalova (Eds.), *Proceedings of the 5th ERME Topic Conference MEDA 2018* (pp. 257–264). Copenhagen, Denmark: University of Copenhagen.

Skott, J. (2015a). The promises, problems, and prospects of research on teachers' beliefs. In H. Fives & M. G. Gill (Eds.), *International Handbook of Research on Teachers' Beliefs* (pp. 13–30). New York: Routledge.

Skott, J. (2015b). Towards a participatory approach to 'beliefs' in mathematics education. In B. Pepin & B. Rösken (Eds.), *From Beliefs to Dynamic Affect Systems in Mathematics Education: Exploring a Mosaic of Relationships and Interactions* (pp. 3–23). Switzerland, Cham: Springer.

Skott, J. (2018). *Re-centring the individual in participatory accounts of professional identity*. In G. Kaiser, H. Forgasz, M. Graven, A. Kuzniak, E. Simmt, & B. Xu (Eds.), *Invited Lectures from the 13th International Congress on Mathematical Education* (pp. 601–618). Switzerland, Cham: Springer.

Skott, J. (2019). Changing experiences of being, becoming, and belonging: Teachers' professional identity revisited. *ZDM Mathematics Education*, 51(3), 469–480. doi:10.1007/s11858-018-1008-3.

Skott, J., Mosvold, R., & Sakonidis, C. (2018). Classroom practice and teachers' knowledge, beliefs, and identity. In T. Dreyfus, M. Artigue, D. Potari, S. Prediger, & K. Ruthven (Eds.). *Developing Research in Mathematics Education – Twenty Years of Communication, Cooperation and Collaboration in Europe* (pp. 162–180). London, New York: Routledge.

Stake, R. E. (2000). Case studies. In N. K. Denzin & Y. Lincoln (Eds.), *Handbook of Qualitative Research* (2nd ed., pp. 435–454). Thousand Oaks: SAGE Publications.

Trouche, L. (2019). Evidencing missing resources of the documentational approach to didactics. Towards ten programs of research/development for enriching this approach. In L. Trouche, G. Gueudet, & B. Pepin (Eds.), *The 'Resources' Approach to Mathematics Education* (pp.447–489). Switzerland, Cham: Springer. doi:10.1007/978-3-030-20393-1.

Wenger, E. (1998). *Communities of Practice: Learning, Meaning, and Identity. Systems Thinker* (vol. 9). Cambridge University Press. https://doi.org/10.2277/0521663636.

INDEX

Page numbers in *Italics* refers figures; **bold** refers table and followed by "n" refers notes number

abstraction in context (AiC) 6, 150–167
abstract-structural mode 140
activity theory 169
Adler, J. 24, 196
"Advice for Action," 72–73
Africa 207
Albano, G. xiv, 2, 3, 7–20
Aldon, G. 69
Amir, O. 104
Anglo-Mexican Spreadsheets Algebra Project 33
apps 26, **27**, 34, 38
artefact 169
artificial intelligence 190
Artigue, M. 150
Asia 207
assessment with technology 70–72, 74–76, 79
Assude, T. 203
audio-podcasts 24, 26, **28**, 29, 38
authentic modelling 153, 154, 157, 164, 165; learning situation 165; problem 154, 165; situations 151; task 164
automatic modelling 156, 159, 162, 163

backwash 100, 103, 108, 111, 114–116
Baldacci, M. 8
Ball, D.L. 192
BAREM platform 97; at macro-level 91–92; at micro-level 92
Basole, R. 86
Battista, M.T. 191
Beck, C. 174
Ben-Naim, D. 104
Biggs, J.B. 100
Black, P. 72, 74, 103
Blumer, H. 216
Blu Tack 145
Bolite-Frant, J. 140
boundary object 139–141
Bruner, J. 141, 142
Brunström, M. xiv, 5, 69–81
Buteau, C. 102, 203
BYJU'S 190

Cabri-Géomètre 198
Cabri Geometry 144
Cain, C. 216
Cambridge Mathematics (CM) Framework 122, 129–133; connections to other resources 132; extensibility and evolution 132; languages 132; multiple maps 130; navigation 132; purpose 130; research summaries 131; structure and relationships 131; unit size 131; waypoints 131
Canadian Mathematics Education Study Group 126
Capobianco, B.M. 104
Carlson, M.P. 151, 155, 156, *156*
CAS *see* Computer Algebra Systems
Casio ClassPad 109

Casyopée software 151, 156–158, 162
Centros de Innovación Educativa Regional (CIER) 43, 48–53, 58n2, **61–68**
Chan Y.C. 145
Charmaz, K 218
Chazan, D. 73, 86, 215
Choppin, J. 84, 169
CIER² (2013–2015) 43, 48–50, 52, 58n2
Clark, L.M. 215
Clark-Wilson, A. xii–xv, 1–6, 121–134
Common Core State Standards in Mathematics (CCSS-M) 122
communities of practice (CoPs) 43, 44, 50, 51, 56, 57
Computadores Para Educar 43, 49, 50, 52, 53, 55, 58n3, **61–68**
Computer Algebra Systems (CAS) 4, 5, 70; alignment 100; backwash effect 100, 108, 111; black-boxing principle 101; CAS-based mathematical activities 102; in Danish secondary school written assessment 107, *107*; dead-end strategies 111; in Denmark 105; design principles 100; epistemic purpose 101; formative assessment 100, 103–104; GeoGebra 109, 110, *110*; in German upper-secondary school written assessment 108–109, *109*; in Germany 105; instrumental genesis 100, 101; instrumentalization 101; instrumentation process 101; intelligent tutoring systems 104; notebooks 106; PISA model 105; pragmatic purpose 101; STACK 113–114; summative assessment 100, 102–103; in United Kingdom 106, 111–112, *113*; whole-class teaching 101
Confrey, J. xv, 3, 4, 121–134
Connected Learning Initiative (CLIx) project 190–192, 196
connected working spaces (CWS) 6, 150–167; analytic description 165; "DG" working space 157; "mathematical functions" working space 158; "measure" working space 157–158; "sheet of paper" working space 157
connectivity 87–96, 206
content knowledge 46, 47, 49, 55, 62
Courtney, S. xv, 2, 4, 6, 84–97
covariation 151, 154–160, 163–165
CPE³ (2014–ongoing) 43, 58n3
criteria 24–28; for quality of task designs 26, 34–36, **35**
curriculum 187–191, 197, 199, 200, 203, 205, 207, 208; development and task design 3–4
DAD *see* documentational approach to didactics
Dalby, D. 104
Deaney, R. 213
de Miranda Quintero, M. 200
dependency 151, 152, 156–161, 163
dependent variable 156, 158, 159, 162–164
de Reuver, M. 86
developing countries 185, 186, 203, 204, 206, 208
didactical tetrahedron 34, 169
Didattica della Fisica e della Matematica (DiFiMa) 43, 48, 49, 52, 53, 58n5
Digital Abilities for All (HDT) programme (2011–2017) 44, 58n9
digital mapping 3, 4, 121, 122, 133; affordances of digital technology 134; Cambridge Mathematics Framework 129–133; challenges 133–134; Dynamic Mathematics Curriculum Network 125–129; future directions 134; learning outcomes 122; Math Mapper 6–8, 122–125
digital platforms 85, 196; contexts 86–87; DAD 87, 88; documentational genesis 87; DRSB, France 86, 87, 91–94, **94**; by educational authorities 85; in grades K–12 85; instrumentalization processes 87, 88; instrumentation 87, 88; macro-level connectivity 88; Meebook, Denmark 86, 92–94, **94**; micro-level connectivity 88; post-secondary education 85; research questions 88–89; by teacher associations 85; Wikiwijs, Netherlands 86, 89–91, **94**
digital resources (DRs) 24, 25, 213; data analysis 218–219; data generation 218; IS initiative 218; leadership 217; Patterns of Participation 214–217; professional identity *see* professional identity; Structuring Features of Classroom Practice 214–215
"Digital Resources for School Database" (DRSB), France 86, 87; BAREM platform at macro-level 91–92; BAREM platform at micro-level 92; macro- and micro-level connectivity 95; macro-level connectivity 94, **94**; platform structure 91; potential instrumentalization processes 95–96
digital textbook 168, 170, 176
dihedral group 143, 144
Dilling, F. xv, 5, 69–81
diversity 198

documentational approach to didactics (DAD) 2, 6, 87–89, 96
documentational genesis 87
documentation work 86–88, 91, 95, 96
Donevska-Todorova, A. xiii, xv, 1–6, 23–39, 138–148
Douady, R. 152
double analysis method 150, 151
Dreyfus, T. 151, 153, 154
Dreyøe, J. xv, 2, 4, 99–116
Drijvers, P. 69, 70
DRs 213–215, 217–226
DRSB *see* "Digital Resources for School Database", France
dual pragmatic-abstract continuum model 138, 141–143, *142*, 147, 148
Duval, R. 152
dynamic geometry (DG) 24, 26, **30**, *31*, 38, 51, 155, 198
dynamic geometry environment (DGE) 24, 70, 71, 99, 139–141, 144; DG added value 31–32; didactical implementation of task 32; instrumental dimension 30–31, *31*; mathematical content 24; pedagogical implementation of task 32
dynamic geometry software 2, 11
Dynamic Mathematics Curriculum (DMC) Network 122, 125; extensibility and evolution 128; interconnected layers of nodes 129, *129*; language selection 127–128; positionality and navigation 128; purpose 127; research participants' model of mathematics concepts and processes 126, *127*; structure and relationships 128; unit size 127

early mathematical computer applications 2
education satellite (Edusat) 187
EMAT 44, 49–51, 54, 58n7, 198
enactive, iconic and symbolic (EIS) representations 141–142
enactive (action-based) mode 141
Enciclomedia 44, 46, 47, 49–54, 58n8, 198, 199
Enseñanza de las Matemáticas con Tecnología (EMAT) 58n7
epistemic modes 101, 110, 116, 139
epistemological continuum 139, 140
estimate 175
estimation 176, 182
e-textbooks 3, 86, 88, 94, 168
European Framework for the Digital Competence of Educators 38–39
European project Intergeo 26
European secondary school 34
experiences of being, becoming and belonging 214, 217, 225
expert teacher 48, 50, 54, 56
explicit educational goals 38
exploratory waypoints 131
Eynon, R. 186

face-to-face learning 3, 48, 51, 53, 58, **64**
Faggiano, E. xiii, xv, 1–6, 42–58
Fahlgren, M. xv, 5, 69–81
feedback 18, 36, 37, 71–76, 78–81, 170, 172, 182, 183
Feldman, A. 104
fifth grade mathematics classroom: analysis 178–182, *179*; data collection 173; estimation tasks 171; instrumental approach 169, 170; *Lehrwerk Mathematik 5*, 171; semiotic mediation 169–170
Fifth Topic Conference of the European Society for Research in Mathematics Education xiii
figured worlds 216–219, 225, 226
focus-group interviews 218
focus of the PD programmes 46–48
Forgasz, H. 203
formative assessment 70–74, 76, 78, 79, 99, 100, 103, 104, 113–115
Formative Assessment in Science and Mathematics Education (FaSMEd) project 72
Forster, P.A. 103
function 150–167
functional dependency 156–158
functional thinking 151
function as covariation 151, 154, 163–165
function modelling 157, 159

Gal, K. 104
game 196
Geoboard 26
GeoGebra files 14, 76, 91, 106, 144, 145, 145, 146, *146*, 148, 189, 191, 192, 203
geographical and cultural diversity of technology integration 3
Geometer's Sketchpad 144
geometric calculation tab 156, 157
geometric reasoning (GR) 190–191
geometry 193
Geraniou, E. xv, 2, 4, 99–116
Ghazawneh, A. 85
Goos, M. 213
Grant, L. 186

graphic calculators 51
Griesemer, J. R. 139
Grønbæk, Niels xiii
Gueraud, V. 104
Gueudet, G. 2, 4, 6, 24
Gueudet, G. xv, 2, 4, 6, 24, 84–97
Gutierrez Rojas, I. 104
Guzmán Guzmán, M.A. 79

Habilidades Digitales para Todos (HDT) 58n9, 200, 201
hands-on activities 191
Hanewald, R. 121
Harlen, W. 100
Häsel-Weide, U. xv, 4, 5, 168–183
Hattie, J. 70, 72, 100
Henfridsson, O. 85
Hennessy, S. 213
Herbst, P. G. 215
Hershkowitz, R. 151
Holland, D. 216
Hoyles, C. 207

iconic (image-based) mode 141
identities 215–217, 225
identity lens 226
identity markers 219, 222–226
identity perspective 226
identity purposes 217
identity studies 216
Ifenthaler, D. 121
independent-dependent variables 162, 163
independent variable 158, 162, 163
India 186–192, 196, 198, 203–206, 208
individualization of teaching and learning 8, 17
information and communication technologies (ICT) 84, 187, 188
infrastructure 198
Innovative School (IS) 213–214, 218
in-service teachers 44, 48, 59
instrumental approach 169
instrumental genesis 55, 100, 101, 115, 140
instrumentalization 88, 89, 91, 95–96
instrumental orchestration 100, 114
instrumentation 86–89, 91–93, 95, 96
instrument-mediated action schemes 169
INTC[1] (1998–2004) 43, 46, 47, 58n1
"intelligent agents," 36
interactive texts (IT) 11; algebraic descriptions 13; elaborating diagrams 13; goal of 11; guiding diagrams 13; interactive diagrams *12*, 12–13; multiple-choice questions 12; online exercises 11; phases 13; preliminary analysis 11; "Putting It All Together," 12
interconnected themes 2
Intergeo European Project 30
internet 47, 49, 52
issues of access 185

Jameson, E. xv, 3, 4, 121–134
Jankvist, U.T. xv, 2, 4, 99–116
Jarvis, D.H. 102
Julie, C. 185, 205, 207

Kafetzopoulos, G.-I. xv, 6, 150–166
Kaiser, G. 151
Karnam, D. xv
Kendal, M. 102
Köcher, N. 75
Koch, M. xv, 3, 4, 121–134
Kock, Z.-J. xv, 2, 4, 6, 84–97
Koehler, M.J. 49
Kondratieva, M. xv, 2, 3, 7–20
Kristinsdóttir, B. xv, 5, 69–81
Kuzniak, A. 152

Laborde, C. 71
Lachicotte, W. Jr. 216
Lagrange, J.-B xv, 6, 150–166
landmark waypoints 131
laptop 201
Latin America 208
Lavicza, Z. 102
learning platforms 190
learning progressions
learning trajectories (LTs) 11, 16, 122–125, *124*, 131, 133
Leung, A. xv, 2, 3, 138–148
lever potential 101, 108, 110, 115
linear sequence of tasks 14
Livne, N.L. 73
Livne, O.E. 73
Logo 59, 197
LogoTurtle 189, 191, 192, *192*, 205
Looney, J. 103

Maier, H. 174
map of gaps 133
Mar, A. 24
Martinez-Maldonado, R. 103
Massive Open Online Courses (MOOCs) 44, 53
mathematical content knowledge 55, **62**

mathematical digital competencies 115
Mathematical Knowledge for Teaching (MKT) model 44
mathematical meaning 170, 183
Mathematical Pedagogical Technology Knowledge (MPTK) model 42, 45–47
mathematical tasks 205, 207, 208
mathematical working spaces (MWS) 152, 166
Mathematics Education in the Digital Age (MEDA 2) 5
Math Mapper 6–8 (MM6–8): learning trajectories 122–125, *124*, 131, 133; positioning nodes and supporting navigation 124–125; purpose of 122–123; structure and relationships 123–124
Mavrikis, M. 104
MCSquared 104
Mead, G. H. 216
meaningful exploration 18
mediation 169
Meebook, Denmark 86, 87; description 92; macro- and micro-level connectivity 93–95, **94**; platform structure 93; potential instrumentalization processes 95–96; potential instrumentation processes 95
Meios INformáticos no Ensino: Racionalização, Valorização, Atualização (MINERVA) 58n10
MERLOT[1] 25
Metafora 104
Mexican-British project 33
Mexican Ministry of Education (SEP) 198
Mexico 186, 198, 199, 201–208
Michener, E.R. 131
middle school 195
migratory approach 71–72, 75, 78, 80
Mindspark 190
MINERVA[10] (1985–1994) 44, 47, 58n10
Minh, T.K. 151, 153, 156
Miranda, L. 200–201
Misfeldt, M. xv, 2, 4, 6, 84–97, 99–116
Mishra, P. 49
model involving quantification 155
modelling 151, 153–157, 159, 161–165; dependencies 156
molecular diffusion in cell *33*, 34
MOOC 44, 48, 53, 57
Moodle 106
Mory, E.H. 73, 171
Mosvold, R. 215
MSWLogo 202
m@t.abel[4] (2005–2012) 43, 46, 48, 58n4
multigrade 201
multimodality 19
multiplication/composition table 146, *147*

Narciss, S. 73
Nardi, E. 7
national digitalization strategy 87
National Education Policy (NEP) 188
national policy discourse 187–188
National Repository of Open Educational Resources 188
networking 150, 165, 166
Noss, R. 116, 207

Ohtani, M. 25
Olsher, S. 73, 75–77
One Laptop Per Child (OLPC) project 189, *189*
Ong, M.Y. 145
online platforms 48
online resources, at university level: aspiration 7; A-T-M perspective 17; A-T-S perspective 17; author's role 8; blended didactic organization 19; blended learning 7; border objects 18; co-action 18; co-disciplinary perspective 8; cognitive dimension 9–10; didactic dimension 10; didactic tetrahedron 8–10, *9*; effective feedback 18; epistemological dimension 10; flexibility and challenge 18; high-quality task design 8; individualization 8, 17–18; interactive texts 11–13, *12*; metacognitive and affective levels 8; multimodality and communication 19; personalization 8; playing metaphor 19; possibility of self-assessment 18; route of activities 13–15; tertiary level of instruction 7; top-level goals 8; Tree of Online Tasks 15–17, *16*, *17*; trial-and-error approach 19; tutor's role 8–9
open educational resources (OER) 86
Organisation for Economic Co-operation and Development (OECD) 186

Pallack, A. 102, 116
Palmer, J.M. 201
Parada, S.E. 200–201
patterns of participation (PoP) 3, 214–219, 225, 226
PD *see* professional development
Peano's axioms 140
Pea, R.D. 71
pedagogical content knowledge 46, 47, 49, **62**, 192

pedagogical technology knowledge 201
Pells, K. 187
Pepin, B. xv, 2, 4, 6, 84–97, 168, 171
personalized and adaptive feedback 4
Phelps, G. 192
Pinkernell, G. xv, 5, 69–81
Plano da Matemática 44
platonic pedagogy 138
"playing" metaphor 19
PLS[6] 43, 58n6
podcasts 34
police quad game 191; classroom dynamics 194; digital prototypes 193, *194*; MIS09 193, *194*; MIS10 193, *195*; paper prototypes 193, *193*
policies 45, 188, 196, 206
Portela, M.J.O. 187
position paper on teaching of mathematics (NCERT) 189
pragmatic 101, 108, 114–116
pragmatic-abstract continuum: abstract-structural mode 140; algebraic and the geometric modes 140, 141; boundary objects 139; concrete-diagrammatic task 145–146; diagrammatic-formal task 146, *147*; dynamic geometry environment 139; enactive, iconic and symbolic representations 141–142; epistemological continuum 139, 140; formal-abstract task 146, *147*; labelling of square vertices 148; pragmatics-concrete task 145; pragmatic weight and abstract weight 141; symmetry of geometric objects 143–145
pragmatics-concrete disk 142
pragmatics-concrete task 145
pragmatic theories 2, 42–60
"pre-algebraic" formulas 158
preservice 199
pre-service teachers 48, 49
PriMaPodcasts 29
primary school 186, 199
private schools 187
procept 142
professional development (PD) 42–60, 188, 201, 205; aspects and sub-aspects 45, **61–68**; "blended" distance programmes 48; CIER[2] (2013–2015) 43, 48–50, 52–53; Colombian PD programmes 50; constraints, allowances and incentives 52–54; craft knowledge 42–43; didactical considerations 47; DiFiMa[5] (2008–ongoing) 43, 48, 49, 52, 53; Digital Abilities for All (HDT) programme (2011–2017) 44; (EMAT, 1998–2007) [7], 44, 49–51, 54; Enciclomedia (2002–2007) 44, 47, 49–53; FUS project 52; in- and out-of-school activities 42; in-service and pre-service teachers 48–49; INTC[1] (1998–2004) 43, 49, 50; integration of theory and practice 49–50; internet connection and technical support 52; MINERVA[10] (1985–1994) 44, 52; MOOCs 44, 53; MPTK model 42, 45; *m@t.abel*[4] (2005–2012) 43, 48, 54; origin and focus 46–48; *Plano da Matemática* (PM; 2006–2012) 44; PLS[6] 43; Portugal's PM 51; *a posteriori* analysis 49; pragmatic theories 42, 55–57; *Programa de Acompanhamento* or Follow-Up of Syllabus (FUS) implementation (1998–2001) 44; resources and materials 51–52; structure of 50–51; teacher enrolment 53; teacher guidebooks 51; Technological Pedagogical Content Knowledge model 49; vision/goals and policies 45–46
professional identity 214, 216, 225, 226; activity 222; analysis and results 223–224; constructivism 219; digital spearhead 221–222, 225; incident 222–223; new and technologically advanced DRs 219–220; routine tasks and guides 221; traditional mathematical teaching 220–221
professional knowledge 200
Progetto m@t.abel 43, 50–52, 54, **61–68**
Programa de Acompanhamento or Follow-Up of Syllabus (FUS) implementation (1998–2001) 44
Programme for International Student Assessment (PISA) 105
Proyecto "Incorporación de Nuevas Tecnologías al Currículo de Matemáticas de la Educación Básica y Media de Colombia (INTC) 43, 45–53, 58n1
Psycharis, G. xv, 3, 6, 150–166, 213–226
pupil-to-teacher ratio (PTR) 187
pushpins 145
puzzle games 144, *144*

quality 24, 26; mathematical technology-enhanced task design 36; resources 25; task designs 23–39; technology-enhanced resources 23–24; technology-enhanced task design 25–26
quantification 155, 162

RA *see* route of activities
Rabardel, P. 169
Rahaman, J. xv, 2, 185–209

RBC-model 153, 166
relational learning clusters (RLCs) 123–124, *124*
replace amplify transform (RAT) framework 80
representative international projects 43–44
resource 24; library 125; quality 25
Restrepo, A. 86
Rezat, S. 4, 5, 168–183
Richard, P.R. 152
Ripley, M. 71
Robert, A. 151
Rocha, H. xv, 42–58
Rocha, K. 2
Rojano, T. xv, 2, 3, 23–39, 185–209
rote learning 187
route of activities (RA) 13; automatic formative feedback 14; Calculus I 13; epistemological and cognitive dimensions 14; feedback 14; for first-year engineering undergraduates 13; GeoGebra 14; methodological-educational cycle 15; multiple representations of subset in plane 14–15, *15*; preliminary analysis 13–14; structure of 14
rural school 189, 190
Ruthven, K. 84, 213, 214

Sabra, H. 86
Sacristán, A. xv, 2, 42–58, 185–212
Sakonidis, C. 215
Salinas-Hernández, U. 201
SAMR model 37
Sangwin, C. 74, 75
Santacruz-Rodríguez, M. xv, 6, 42–58
Schacht, F. xv, 4, 5, 168–183
Schmidt, W.H. 122, 130
Schreiber, C., xv, 3, 23–39
Schwarz, B.B. 151
Sebastian, R. xv
Seeing The Entire Picture (STEP), Israel 69, 75–77, 79, 80; aspects of feedback 77; description 75; formative and summative assessments 76; functionalities of technology 76–77; migratory and transformative 76; with and through assessments 76
self-assessment 18
semiotic dimension 157, 158, 161, 163–165
semiotic mediation 140, 169, 173
SFCP *see* Structuring Features of Classroom Practice
Shaughnessy, J.M. 193
Shute, V.J. 73
Siegert, B. 121, 125
Sinclair, N. 84
Skinner, D. 216
Skott, Charlotte Krog xv, 3, 6, 213–226
Skott, J. xv, 3, 6, 213–226
SMART *see* Specific Mathematics Assessments that Reveal Thinking, Australia
smartboards 199
social construction of meaning 181
social interaction 170, 173
social practice perspective (PoP) 6
social practice theory 214, 216
socio-economic divide 186, 187
Sørensen, C. 86
Specific Mathematics Assessments that Reveal Thinking (SMART), Australia 5, 69, 77–81; aspects of feedback 78–79; description 77; formative and summative assessment 78; functionalities of technology 78; guidelines 78; migratory and transformative assessment 78; test questions 77; with and through assessment 78
spreadsheets 24, 32–34, 38, 51, 198
Srinivas, S. xv, 2, 185–209
Stacey, K. 70, 77, 78, 80, 102
STACK *see* System for Teaching and Assessment using a Computer algebra Kernel, United Kingdom
Star, L.S. 139
STEP *see* Seeing The Entire Picture, Israel
strategies and methods of the PD programmes 48–52
Structuring Features of Classroom Practice (SFCP) 213–215, 219, 224, 226
student learning apps 190
student motivation 187
styrofoam board 145
substitution, augmentation, modification, redefinition (SAMR) model 36–37
summative assessment 70, 71, 73, 74, 76, 78, 79, 100, 102–103, 106, 108, 111, 113, 115, 116
Sutherland R. 204
Suurtamm, C. xv, 3, 4, 103, 121–134
Swan, M. 104, 131
symbolic (language-based) mode 141
symmetry group 143
symmetry of geometric objects: primary school level 144–145; secondary school level 144; university level 143
symmetry of the square 145
systematical-extensional analysis 174

System for Teaching and Assessment using a Computer algebra Kernel (STACK), United Kingdom 5, 69, 74–75, 79, 113–114; aspects of feedback 75; CAS-based analysis 74; description 74; formative and summative assessment 74; functionalities of technology 75; JAVA-based interactive elements 74; migratory and transformative assessment 75; with and through assessment 75

tablet applications 27, 28
Tall, D. 141, 193
Tamborg, A. xv, 2, 4, 6, 84–97
Tang, C.S.-K. 100
task-formats 25, 27
tasks 25, 27, 170
Tata Trust Report 189
Teacher: identity 216; knowledge 200; needs 48, 56
Teaching Mathematics with Technology (EMAT) program 198–200
technology added value 36
technology availability 204
technology-enhanced resources 23–24, 26, 34–36; audio-podcasts for primary schools 24, 28–30; Australian project 26; DGE for secondary schools 24, 30–32, *31*; digital resources 24, 25; dynamic geometry environments for secondary schools 24, 30–32, *31*; European project Intergeo 26; Geoboard 26; non-digital teaching and learning environments 25; quality in education 24–25; quality of task design 34–36, **35**, **37**, 38; research questions and research methods 26–27; SAMR model 36–37; spreadsheets for upper secondary schools 24, 32–34, *33*; tablet apps for primary school 27–28; verb resource 24; video- or audio-resources 26
technology integration, in India 185; access in schools 205–206; classroom dynamics 196–197; curricular/task implementation 207; (paper-and-pencil) examination performance 187; implementation in classroom 191–192, *192*; infrastructural and situational issues 197; interrelated factors 203–204; Mindspark 195–196; opportunities 197–198; police quad game 193–194, *193–195*; policy issues 206–207; in primary and secondary schools 187–189, *188*, *189*; pupil-to-teacher ratio 187; in rural schools 190–191; teacher acceptance and adoption 197; technology availability issues 204–205
technology integration, in Mexico: access in schools 205–206; curricular/task implementation 207; in government primary and secondary schools 198–200; implementation issues 200–201, *202*; interrelated factors 203–204; policy issues 206–207; rural multigrade primary school 202–203; technology availability issues 204–205
technology-rich assessment 5; assessment with and through technology 70–71; FaSMEd project 72; feedback 72–73; "mathematically able" software 70; migratory and transformative approach 71–72; paper-based assessment 70; RAT framework 80; SMART, Australia 69, 77–79; STACK, United Kingdom 74–75; STEP, Israel 69, 75–77; summative and formative assessment 70
Telesecundaria 198
television 198
Telloni, A.I. xv, 2, 3, 7–20
tessellations 144, *144*
tetrahedron model 3, 8
Thames, M.H. 192
third-grade mathematics classroom: analysis 174–177; data collection 172–173; *Denken und Rechnen interaktiv* 170–171; estimation tasks 171–172, *172*, *173*; instrumental approach 169, 170; semiotic mediation 169–170
Thomas, M.O.J. 201
Thompson, P.W. 151, 155
TI-92 algebraic calculator 198
Timperley, H. 70, 72
TISS 191, 193, 194
Tondeur, J. 44
transformative approach 71–72, 75, 79, 80
tree of online tasks 15
Tree of Online Tasks (TT) 15–17, *16*, *17*
Trgalova, J. xiii, xv, 1–6, 23–39
Trouche, L. 24, 32, 44, 86, 87, 91, 96, 198, 200, 201, 207, 214
TurtleArt 189
Turtle Blocks 189
turtle geometry 203

UNICEF 206
university transitions 8
upper-secondary level: abstraction in context 153, 158–159; activity sequence 156;

choice of digital environment 156; combining AiC and CWS 153–154; connected working spaces 152–153, 157–166; context 154; distinction between independent-dependent variables 162–163; expressing covariation through the use of variables 163–164; extra-mathematical situations 151; gutter problem 154–155, *155*; identifying dependencies through paper model 160; modelling in DG 161; quantification of variations 162–163; task presented to students 157
usage schemes 169, 170, 176, 177

Vandebrouck, F. 151
variables 156, 158, 159, 162–164
vertical mathematization 153
Vinner, S. 193

Watson, A. 25
waypoints 131
Weigand, H.-G xv, xiii, 1–6, 69–81, 99–116
Wenger, E. 216
Wight, C.A. 73
Wikiwijs, Netherlands 86; macro- and micro-level connectivity 94, **94**, 95; macro-level constraint 90; micro-level constraint 90–91; platform structure 89–90; potential instrumentalization processes 95–96; potential instrumentation processes 95
Wiliam, D. 70, 72, 74, 103
Wollring, B. 25

Yerushalmy, M. 73, 75, 76, 86
Youngsters' Use of Digital Media 214, 215, 218